Robert C

W9-BMU-988

AC Theory

NJATC

AC Theory

NJATC

THOMSON

DELMAR LEARNING

Australia   Canada   Mexico   Singapore   Spain   United Kingdom   United States

THOMSON

DELMAR LEARNING

AC Theory
NJATC

Vice President, Technology and Trades SBU:
Alar Elken

Executive Director, Professional Business Unit:
Gregory L. Clayton

Product Development Manager:
Patrick Kane

Development Editor:
Angie Davis

Channel Manager:
Beth A. Lutz

Marketing Specialist:
Brian McGrath

Production Director:
Mary Ellen Black

Production Manager:
Larry Main

Senior Project Editor:
Christopher Chien

Art/Design Coordinator:
Francis Hogan

For permission to use material from the text or product, contact us by
Tel.   (800) 730-2214
Fax   (800) 730-2215
www.thomsonrights.com

Library of Congress Cataloging-in-Publication Data:

NJATC.
    AC Theory / NJATC.
       p. cm.
    ISBN 1-4018-5685-3
    1. Electric circuits—Alternating current.
    I. Title.
    TK454 .15.A48 A25 2004
    621.3815′2—dc22
                    2004000593

NOTICE TO THE READER

Contents

CHAPTER 2
Mathematics, Using Vectors Effectively   22

CHAPTER 3
Comparing AC to DC

CHAPTER 4
AC Resistive Circuits

CHAPTER 5

Three-Phase Systems

58

PART 2

AC AND DC GENERATORS

69

CHAPTER 6

DC Generators

70

CHAPTER 7

AC Generators

86

PART 3
INDUCTANCE IN AC CIRCUITS   99

CHAPTER 8
Inductance and Its Effects on Circuits   100

CHAPTER **9**
Inductive Reactance   **118**

CHAPTER **10**
Inductors in Series and/or Parallel   **126**

PART 4
CAPACITANCE IN AC CIRCUITS   135

CHAPTER **11**
Capacitance and Its Effect on Circuits   136

CHAPTER **12**
Capacitive Reactance   150

CHAPTER 13

Capacitors in Series and/or Parallel   158

PART 5

COMBINING RESISTANCE, INDUCTANCE, AND CAPACITANCE   169

CHAPTER 14

Characteristics of AC Circuits   170

CHAPTER 15
Parameters of Series RL Circuits   184

CHAPTER **16**
Parallel RL Circuits   **196**

CHAPTER **17**
Series RC Circuits   **204**

CHAPTER 18
Parallel RC Circuits
212

CHAPTER 19
Inductive and Capacitive (LC) Circuits   220

CHAPTER 20
Resistive, Inductive, and Capacitive (RLC) Circuits in Series   234

CHAPTER 21

Resistive, Inductive, and Capacitive (RLC) Circuits in Parallel

242

CHAPTER **22**

Comparing Resistive, Inductive, and Capacitive (RLC) Circuits in Series and Parallel 252

CHAPTER **23**
Combination RLC Circuits   **266**

CHAPTER 24
Series RLC Circuits and Resonance   274

CHAPTER 25
Parallel RLC Circuits and Resonance   284

PART 6

ADDITIONAL AC TOPICS   293

CHAPTER 26
Use of Filters to Control AC Signals   294

CHAPTER **27**
Power Factors in AC Circuits   312

Preface

The National Joint Apprenticeship and Training Committee (NJATC) is the training arm of the International Brotherhood of Electrical Workers and National Electrical Contractors Association. Established in 1941, the NJATC has developed uniform standards that are used nationwide to train thousands of qualified men and women for demanding and rewarding careers in the electrical and telecommunications industries. To enhance the effectiveness of this mission the NJATC has partnered with Delmar Learning to deliver the very finest in training materials for the electrical profession.

Knowledge of fundamentals is critical to the success of a modern electrical technologist. Every project, every piece of knowledge, and every new task will be based on all of the experience and information that you get as you progress through your career. This book contains much of the material that will form the foundation of your electrical knowledge.

Starting with the methods and techniques that you have learned in your study of DC circuit theory, this book adds the concepts of capacitance and inductance as they relate to alternating current (AC) theory. In addition to presenting the fundamentals of capacitance and inductance, this book also provides a variety of circuit analysis tools that are extended versions of the concepts you have already learned.

Like the DC Theory book that is part of this series, this AC Theory book is divided into six parts to help structure your learning. Each part contains chapters that progress in a rational, well-paced schedule that enables you to become thoroughly familiar with one set of principles before you move on and apply your new knowledge to learning more advanced topics.

PART 1 INTRODUCTION TO AC THEORY

Chapter 1 provides a brief review of some of the more important fundamentals of *DC theory*. You will review electron theory, Ohm's law, and other such baseline knowledge that is so important to the electrician.

Chapter 2 introduces you to the concept of vectors and complex numbers. These critical mathematical tools are a must for anyone involved in circuit analysis of circuits that contain resistive, inductive, and capacitive elements. Usually considered to be "advanced math," the topics covered in this chapter are presented in a straightforward, practical manner.

Chapters 3 and 4 start your immersion into AC by comparing it to DC. Various parameters that you must know to understand and analyze AC circuits are introduced by comparing them to already familiar DC concepts. Additionally, you will learn that when only resistive elements are present AC circuit analysis is identical to DC circuit analysis. The chapters provide examples and practice problems that enable you to apply beginning AC principles.

Chapter 5 completes your introduction to AC circuits by presenting the concept of polyphase circuits. Since virtually all major power distribution systems throughout the world use polyphase circuits, knowledge of polyphase principles is critical.

PART 2 AC & DC GENERATORS

This part (Chapters 6 and 7) reintroduces the basic principles of AC and DC generators by highlighting the differences and similarities between the two. The modern electrician can hardly expect to function in modern power systems without a complete understanding of the critical information presented in this part.

PARTS 3 AND 4 INDUCTANCE AND CAPACITANCE IN AC CIRCUITS

The six chapters in these two parts introduce the second and third elements of the passive electrical circuit—capacitors and inductors. Here you will learn the fundamental operating principles of each, how they effect circuit voltages and currents, and how to account for their effects when they are connected in series and/or parallel.

PART 5 COMBINING RESISTANCE, INDUCTANCE, AND CAPACITANCE

This is the largest part of the text and comprises Chapters 14 through 25. In this part you will learn how to analyze all of the possible combinations of RLC circuits including series, parallel, and combination circuits. You will learn that such calculations are very similar to simple resistance calculations with the requirement that vectors (complex numbers) must be used in the mathematical calculations. The methods used do not emphasize mathematical techniques. Rather you are encouraged to use simple tools such as the rules of a right triangle and simple addition and subtraction to perform your calculations.

PART 6 ADDITIONAL AC TOPICS

The final part introduces two practical applications of the material that you have learned in the previous 25 chapters. Chapter 26 introduces the concept of electrical filters which are used to eliminate and/or pass certain frequencies. Filters are found throughout the electrical industry in every application from electronics to 765 kV power systems.

The final chapter focuses on the application of AC circuit theory to that of power factor, the ratio of "real power" to "apparent power." For those readers who work in electrical power distribution systems, the concept of power factor is one that you will work with again and again throughout your career.

The NJATC can provide a complete line of electrical and telecommunication training materials, including CBT programs and courses. Visit the NJATC online at njatc.org to review the finest electrical training curriculum the industry has to offer. The subject of Alternating Current is both interesting and essential for the electrical and electronic student. Take the time to progress through the AC text material, perform the calculations and review the chapter objectives before moving forward to the next section. Your understanding of AC Theory will provide all of the essentials to move to the next level of expertise in the electrical and electronic fields. Should you decide on a career in the electrical industry, the International Brotherhood of Electrical Workers and the National Electrical Contractors Association (IBEW-NECA) training programs provide the finest electrical apprenticeship programs the industry has to offer. If you are accepted into one of their local apprenticeship programs you'll be trained for one of four career specialties, journeyman lineman, residential wireman, journeyman wireman or VDV installer/technician. Most importantly, you'll be paid while you learn. To learn more visit http:/www.njatc.org.

NJATC ACKNOWLEDGEMENTS
Principal Writer
Stan Klein, NJATC Staff

Contributing Writer
John McCord, Instructor, Vineland, NJ

ADDITIONAL ACKNOWLEDGEMENTS

This material is continually reviewed and evaluated by Training Directors who are also members of the NJATC Inside and Outside Education Committees. The invaluable input provided by these individuals allows for the development of instructional material that is of the absolute highest quality. At the time of this printing the Education Committee was comprised of the following members:

Inside Education Committee

Dennis Anthony—Phoenix, AZ; John Biondi—Vineland, NJ; Dan Campbell—Tangent, OR; Peter Dulcich—Syracuse, NY; John Gray—San Antonio, TX; Gary Hunziker—Sacramento, CA; Dave Kingery—Salt Lake City, UT; Bill Leigers—Richmond, VA; Bud McDannel—West Frankfort, IL; Bill McGinnis—Wichita, KS; Jerry Melson—Bakersfield,

CA; Tom Minder—Fairbanks, AK; Bill Newlin—Dayton, OH; Jim Paladino—Omaha, NE; Dan Sellers—Collegeville, PA; and Jim Sullivan—Winter Park, FL.

Outside Education Committee

Charley Young—Lawrence, KS; SK Pelch—Sandy, UT; Armando Mendez—Riverside, CA; Steve Uhl—Limerick, PA; Bill Stone—Portland, OR; Don Jamison—Indianola, IA; Howard Miller —Medway, OH and Virgil Melton—Atlanta, GA.

PUBLISHER ACKNOWLEDGEMENTS

John Cadick, P.E., Contributor

A registered professional engineer, John Cadick has specialized for almost four decades in electrical engineering, training, and management. In 1986 he founded Cadick Professional Services (forerunner to the present-day Cadick Corporation), a consulting firm in Garland, Texas. His firm specializes in electrical engineering and training, working extensively in the areas of power system design and engineering studies, condition-based maintenance programs, and electrical safety. Prior to the creation of Cadick Corporation, John held a number of technical and managerial positions with electric utilities, electrical testing firms, and consulting firms. Mr. Cadick is a widely published author of numerous articles and technical papers. He is the author of the Electrical Safety Handbook as well as Cables and Wiring. His expertise in electrical engineering as well as electrical maintenance and testing coupled with his extensive experience in the electrical power industry makes Mr. Cadick a highly respected and sought after consultant in the industry.

Monica Ohlinger

The publisher would like to thank Monica Ohlinger of Ohlinger Publishing Services for her diligent work in development on the text.

PART

1

INTRODUCTION TO AC THEORY

chapter 1

Reviewing the Applications of DC Theory

■ OUTLINE

OVERVIEW

Alternating current, usually abbreviated as AC, is used to provide over 90% of the world's electrical power requirements. In the very early years of electrical power, direct current (DC) was the most widely used form of electricity. The beginnings of the transition from a DC world to an AC world were turbulent and full of technical and political disagreements and intrigue. Thomas Edison and George Westinghouse, outspoken competitors in those early years, staked their reputations and their businesses on their particular systems.

Eventually, for a variety of technical and business reasons, Westinghouse won the dispute. His AC transformers and motors, developed by the brilliant technical genius Nikola Tesla, proved the edge that overcame Edison's DC systems. Despite this fairly early victory for AC, at least two large American cities continued to use DC distribution in their downtown areas until the 1970s.

This is the beginning of your study of AC theory. Your success in understanding the material presented in these chapters is dependent on how well you understand the basic principles of DC theory. The concepts learned in your earlier study of DC theory will also apply to your study of AC.

This chapter reviews basic DC theory and formulas and requires some calculations. After successfully completing this chapter, you should have no problem continuing your study of AC theory. Should you encounter any problems with this chapter, you should review the earlier material.

OBJECTIVES

After completing this chapter, you should be able to:

1. Explain the basic construction of the atom and how it relates to electrical theory.
2. Describe the effects that conductor material and construction have on resistance.
3. Use basic circuit analysis tools to analyze DC series, parallel, and combination circuits. The analysis tools include Ohm's law, the superposition theorem, Kirchhoff's laws, Thevenin's theorem, and Norton's theorem.
4. Explain the theory and operation of magnetism, electromagnetism, induction of voltage with magnetism, and AC and DC generators.

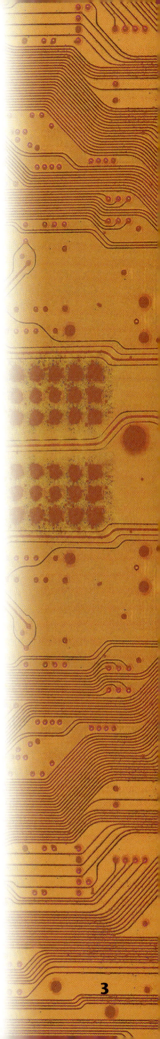

■ GLOSSARY

Acid   Any of a large class of sour-tasting substances whose aqueous solutions are capable of turning blue litmus indicators red, of reacting with and dissolving certain metals to form salts, and of reacting with bases or alkalis to form salts. A substance that ionizes in solution to give the positive ion of the solvent. A substance capable of yielding hydrogen ions. A proton donor. An electron acceptor. A molecule or ion that can combine with another by forming a covalent bond with two electrons of the other.[1]

Alkali   See Base.

Alternating current (AC)   A current that varies, or "alternates," from one polarity to another.

Ampere   The measure of a specific number of electrons that pass a specific point in 1 second. That number is approximately 6.25×10^{18}, or 6.25 billion billion, electrons and is called a *coulomb.* When that number of electrons passes a specific point in 1 second, we say that 1 amp is flowing and is represented by the symbol **A.** In calculations, current is represented by the letter "I."

Anion   A negative ion.

Base   Any of a large class of compounds, including the hydroxides and oxides of metals, having a bitter taste, a slippery solution, the ability to turn litmus blue, and the ability to react with acids to form salts. A molecular or ionic substance capable of combining with a proton to form a new substance. A substance that provides a pair of electrons for a covalent bond with an acid. In *electronics,* the region in a transistor between the emitter and the collector; the electrode attached to this region.[2]

Cation   A positive ion.

Conductor   A material whose electrons can be moved with relative ease.

Direct current (DC)   A current that flows in one direction only.

Electrolyte   A liquid or semiliquid material created by dissolving metallic salts in solution. An electrolyte is a conductor.

Electromotive force   The electrical pressure generated between two areas with different amounts of electrical charge.

Element   A substance composed of atoms having an identical number of protons in each nucleus. Elements cannot be reduced to simpler substances by normal chemical means.[3]

Insulator   A material whose electrons strongly oppose movement.

Ion   Any particle with an electrical charge. Usually the term refers to atoms or molecules that have either gained or lost electrons.

Isotope   An atom of an element with a different number of neutrons. For example, the normal hydrogen atom has one electron and one proton. The isotope of hydrogen called "deuterium" has one electron, one proton, and one neutron. Different isotopes of the same element are chemically indistinguishable from each other.

Ohm   The unit of resistance in a circuit. Specifically, it is the amount of resistance that allows 1 ampere of current to flow when 1 volt is applied. The symbol used to represent the ohm is the Greek letter omega (Ω). In calculations, resistance is represented by the letter "R." A component in a circuit that creates resistance is called a *resistor.*

Periodic Table of the Elements   A tabular arrangement of the elements according to their atomic numbers so that elements with similar properties are in the same column.[4]

Piezoelectric   Generation of electricity from pressure and vice versa.

Polarity   Opposing or opposite characteristics. For example, the electron has a negative charge, and the proton has a positive charge. The polarity of the electron is said to be negative, and the polarity of the proton is positive.

Semiconductor   A material with four electrons in the valence shell. Semiconductors have more electrical resistance than conductors but fewer than insulators.

Valence shell   The outermost shell, or orbit, of electrons in an atom

Volt   The electromotive force that pushes electrons through the conductors, wires, or components of a circuit. It is similar to the pressure exerted on a system of fluid using pipes. The higher the pressure, the more flow. Specifically, the volt is the amount of work done per coulomb of charge (volts = joules per coulomb) and is represented by the symbol **V.** In calculations, voltage is represented by the letter "E." Remember that voltage is the force required in creating flow, but volts do not flow through the circuit.

---

[1]Excerpted from *American Heritage Talking Dictionary.* Copyright © 1997 The Learning Company, Inc. All Rights Reserved.
[2]Excerpted from *American Heritage Talking Dictionary.* Copyright © 1997 The Learning Company, Inc. All Rights Reserved.
[3]Excerpted from *American Heritage Talking Dictionary.* Copyright © 1997 The Learning Company, Inc. All Rights Reserved.
[4]Excerpted from *American Heritage Talking Dictionary.* Copyright © 1997 The Learning Company, Inc. All Rights Reserved.

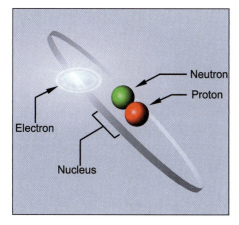

FIGURE 1–1   The hydrogen atom (deuterium).

■ ELECTRON THEORY

Electricity is an invisible force that can produce heat, motion, light, and any number of other physical effects. This invisible "driving force" provides power for lightning, radios, motors, the heating and cooling of buildings, and many other applications. The common link among all of these applications is the electrical charge. All the materials we know—gases, liquids, and solids—contain two basic particles of electrical charge: the electron and the proton. The electron has an electrical charge with a negative **polarity**. The proton has an electrical charge with a positive polarity.

Atoms have three main parts: electron, proton, and neutron. The proton and neutron combine to form the atom's nucleus. Hydrogen, the simplest of all **elements**, has a single proton, and a single electron. Figure 1–1 is a diagram of an **isotope** of hydrogen called deuterium. Deuterium has one electron, one proton, and one neutron. Recall that the electron has a negative (−) charge and the proton a positive (+) charge. The neutron, as the name suggests, has no electrical charge. In other words, the neutron is electrically neutral and has no effect on the electrical characteristics of the material.

You can tell the type of an element by the number of protons in the atom's nucleus. For example, silver has 47 protons in its nucleus, iron 26, and oxygen 8. The number of protons also equals the element's atomic number in the **Periodic Table of the Elements**. Although there are many possible ways in which electrons and protons might be grouped in atoms, they come together in very specific combinations that produce stable arrangements (atoms). The simplest element is hydrogen and its isotopes.

Figure 1–2 shows a copper atom with a number of rings filled with electrons in orbit around the atom's nucleus. There are many orbital rings (shells) around a nucleus that can hold electrons.

1.1 Electron Flow

The electrons in the outermost shell of some materials can flow from one atom to the next. The flow of electrons is always from the pole of an electrical energy source that has an excess of electrons to the pole that has a deficiency of electrons. The direction of current flow, negative to positive, is defined as the polarity of the current flow. Figure 1–3 shows how this electron flow works.

Materials whose electrons have a very high opposition to being moved are called **insulators**. These materials cannot conduct very well because their electrons do not readily move from atom to atom. Materials with atoms in which the electrons have a very low opposition to being moved are known as **conductors**.

Other materials are called **semiconductors**. Semiconductors pass electrons less easily than conductors but more easily than insulators. Semiconductors have four electrons in the outermost ring (the **valence shell**). When heated, the semiconductor's resistance to electron flow decreases. A conductor's resistance to electron flow increases when heated.

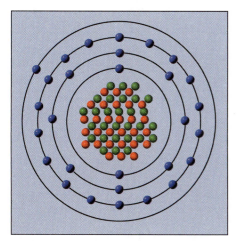

FIGURE 1–2   Electrons in their orbits (copper atom).

FIGURE 1–3   Electron flow.

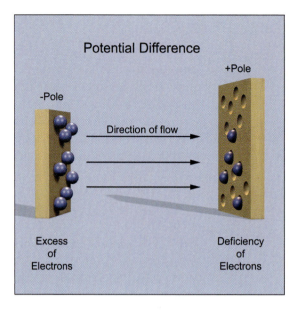

1.2 Ions

Ions are particles that have an electrical charge. An atom that has lost electrons is considered a positive ion, called a **cation**. An atom that has gained electrons is considered a negative ion, called an **anion**. The process of an atom losing or gaining electrons is called *ionization*. When atoms of different elements exchange electrons this way, compounds are formed. For example, when sodium (a metal) is combined with certain active nonmetals (e.g., chlorine), salts, such as sodium chloride (Na^+Cl^-), are formed. These ionized salts can be mixed with water (a nonconductor) to form a conducting liquid (see Figure 1–4).

Acids and **alkalis** (also called **bases**), along with these types of "metal" salts, are also used to cause electrical conduction in liquids. These types of solutions are called **electrolytes**. For example, sulfuric acid is made up of three ions (hydrogen, sulfur, and oxygen) and is the

FIGURE 1–4   Electrolytic action.

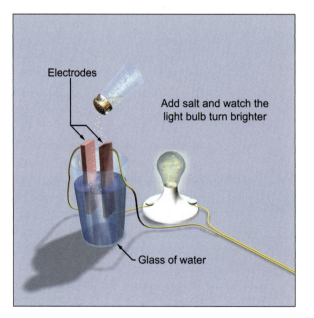

main electrolyte (electrolytic liquid) used in lead-acid car batteries. Sulfuric acid's chemical symbol is H_2SO_4. This indicates that sulfuric acid is made up of two hydrogen atoms, one sulfur atom, and four oxygen atoms.

1.3 Sources of Electrical Energy

There are six methods that are known to force electrons from the valence ring of an atom to become a potential electrical current participant:

1. Friction
2. Chemical
3. Heat
4. Pressure
5. Light
6. Magnetism

Except for friction, electrical energy can produce the same effects as those used to create it. That is, if the proper load is connected across a source, any one (or more) of the following effects will occur:

1. Chemical action
2. Heat
3. Pressure
4. Light
5. Magnetic fields

The chemical source of electricity is best represented in our everyday lives by the battery. The battery is a primary source of the **direct current (DC)**. Batteries are usually divided into two categories: primary cells and secondary cells. Primary cells, like the batteries used in portable radios and flashlights, cannot be recharged. Once they have depleted their chemical action, they must be thrown away. Secondary cells, like those used in automobiles, can be recharged several times.

Heat is generated whenever current flows through a wire. This happens because energy is used to move electrons. Usually, the effect of heating a wire is undesirable. For certain applications, heat can be a desirable outcome. A transfer of electrons can also take place when two dissimilar metals are joined together at a junction and then heated. This is known as the *thermoelectricity process.* Increases in the temperature (heat) cause a greater transfer of electrons. This type of device is called a *thermocouple.*

Certain crystalline substances, when placed under pressure, generate minute **electromotive forces**. These forces cause the electrons to be driven out of orbit in the direction of the force. The electrons leave one side of the material and collect on the other side. Electricity derived from pressure is known as the **piezoelectric** effect. It is possible to reverse the piezoelectric process and produce pressure with electrical current. This principle is used in some applications, such as in very small earphones that use a piezoelectric crystal to produce the sound vibrations from an electrical signal.

Light energy can produce electricity, and electricity can produce light. It works both ways. Light is made up of small particles of energy called *photons.* When photons strike certain types of photosensitive material, they release energy into the material. There are three types of photoelectric effects of interest in the study of electricity: photo-emission, photovoltaic, and photoconduction. Electrical current produces light if enough current is passed through a poor conductor. In this example, not only is heat generated, but many materials will begin to glow red or even white hot, as in an incandescent lamp. There are three other methods of producing light with electricity that do not result in as much heat loss: electroluminescence, phosphorescence, and fluorescence.

Magnetism is the primary source of electrical power. Whenever current flows through a conductor, heat and a magnetic field are generated. When a conductor (a material in which the valence electrons can be easily removed from their orbit) moves through a magnetic field, electrons will move from one end of the conductor to the opposite end, creating a potential source between the two ends of the conductor. This process of producing electrical energy is known as *magnetoelectricity* and makes electrical devices such as electrical magnets, motors, and transformers possible. Magnetoelectricity is the basis for commercial generators. Most electrical power is generated by this method.

■ OHM'S LAW

Ohm's law is a law of electrical proportionality. It states that 1 **volt** will push 1 **ampere** through 1 **ohm** of resistance. Another way to look at this relationship is that the current (*I*) in amperes is directly proportional to the voltage (*E*) in volts and inversely proportional to the resistance (*R*) in ohms. Ohm's law can be expressed mathematically as

$$E = IR \ or \ I = \frac{E}{R} \ or \ R = \frac{E}{I} \qquad (1.1)$$

EXAMPLE 1

A circuit has a voltage of 24 V and has a current of 2 A. What is the resistance of the circuit?

Solution:
Using Ohm's law, *R* becomes

$$R = \frac{E}{I} = \frac{24 \text{ V}}{2 \text{ A}} = 12 \text{ }\Omega$$

1.4 Cross-Sectional Areas and Resistance

The cross-sectional area of a material has a great effect on the resistive characteristics of any conductor using that material. As conductors are made larger in diameter, the number of atoms contributing free electrons increases and tends to reduce the resistance for that conductor. For example, a #6-AWG copper wire has a diameter of 0.162 inches; therefore, the cross-sectional area is

$$A = \pi r^2 = \pi \left(\frac{d}{2}\right)^2 = 3.14159 \times \left(\frac{0.162}{2}\right)^2 = .00651 \text{ in.}^2 \quad (1.2)$$

However, because of the inconvenience of working with such small numbers, most calculations are done using the circular-mil area of a conductor. A mil is defined as 1/1,000 of an inch, or 1,000 mils equal 1 inch. To convert from wire diameters that are expressed in inches to wire diameters in mils, simply multiply the number by 1,000. The #6 wire mentioned earlier would have a diameter of 0.162 × 1,000, or 162 mils.

The circular-mil area of a conductor is defined as the diameter of the conductor in mils squared. The circular-mil (abbreviated as cmil) area can be found by simply taking the diameter of the wire in mils times itself. The #6 wire would have a circular-mil area of A = 162 × 162 = 26,244 circular-mils (cmils). Figure 1–5 shows two examples of cross-sectional areas and diameters for wires.

1.5 Ohm's Law and Power

The unit of electrical power is the watt. Power in a circuit is the amount of work being done per unit time. The load in a circuit uses the energy at a certain rate called power. Power, in equation form, looks like this:

$$\text{Power} = \frac{\text{Work}}{\text{Time}} \quad (1.3)$$

The work in this equation is the force applied to the circuit times the distance that the force moves something. In an electrical circuit, the

FIGURE 1–5 Wire diameters in circular mils (cmils).

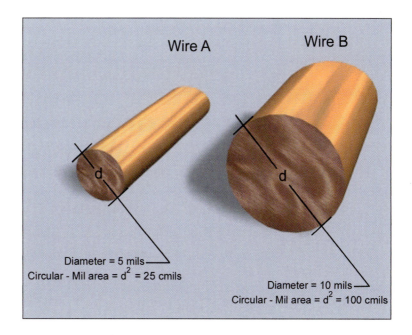

Wire A

Wire B

d

d

Diameter = 5 mils
Circular - Mil area = d² = 25 cmils

Diameter = 10 mils
Circular - Mil area = d² = 100 cmils

power is the volts (joules/coulomb) times the amps (coulombs/second) and equals watts (joules/second). Another way to look at power is the amount of energy used per second. Mathematically, the heat or power produced can be shown as

$$P = I^2R \tag{1.4}$$

The heat produced by the current through the resistance in the circuit is called "I^2R" losses because it is heat lost in the system. Two other formulas can be found using variations of the $P = I^2R$ equation:

$$P = \frac{E^2}{R} \text{ and } P = EI \tag{1.5}$$

EXAMPLE 2

What is the power loss due to heat in a motor that draws 20 A of current and has a resistance of 5 Ω?

Solution:
Since we know current and resistance, the formula to use is $P = I^2R$:

$$P = I^2R = (20A)^2 \times 5 \ \Omega = 2,000 \ W$$

EXAMPLE 3

A room heater has a rating of 2,000 W and uses 120 V to power it. What is the resistance of the heater?

Solution:
Since we know the power and the voltage, the formula to use is $P = E^2/R \text{ or } R = \dfrac{E^2}{P}$:

$$R = \frac{E^2}{P} = \frac{(120 \ V)^2}{2,000 \ W} = 7.2 \ \Omega$$

■ DC CIRCUIT ANALYSIS

There are three basic types of DC circuits: series, parallel, and compound (a combination of series and parallel).

1.6 Series Circuits

The total resistance in a series circuit is the sum of the individual resistors and can be calculated by the following equation:

$$R_T = R_1 + R_2 + R_3 + \ldots R_N \tag{1.6}$$

where R_T = total resistance.

Since all the series circuit current flows through every component in the circuit, each component represents a load (resistive element) in determining the total current. Ohm's law defines this relationship between voltage, resistance, and current as

$$I_1 = I_2 = I_3 = I_T \text{ and } I_T = \frac{E_T}{R_T} \qquad (1.7)$$

where I_T = total circuit current.

The total voltage source in a series circuit is the algebraic sum of the individual voltage source:

$$E_T = E_1 + E_2 + E_3 + \ldots E_N \qquad (1.8)$$

Remember to keep the polarity of each voltage source. Voltages with opposite polarities will subtract from each other.

Like E_T and R_T in a DC circuit, the power consumed by the entire circuit is the sum of the power used by each component:

$$P_T = P_1 + P_2 + P_3 + P_4 + \ldots P_N \qquad (1.9)$$

1.7 Parallel Circuits

Parallel circuits are ones that have more than one path for current to flow. These different paths are called *parallel branches.* The voltage applied to each of the parallel branches is equal to the voltage across the branch. The sum of the branch currents is equal to the total current. Mathematically,

$$E_T = E_1 = E_2 = E_3 = \ldots E_N \qquad (1.10)$$

and

$$I_T = I_1 + I_2 + I_3 + \ldots I_N \qquad (1.11)$$

where N is the number of branches in the parallel circuit.

The total resistance of a parallel circuit is always less than the smallest resistance of any one branch. To calculate the total resistance in a parallel circuit, the so-called reciprocal formula is used. The reciprocal equation was developed in the *DC Theory* text and can take either one of two equivalent forms:

$$\frac{1}{R_T} = \frac{1}{R_1} + \frac{1}{R_2} + \frac{1}{R_3} + \ldots \frac{1}{R_N} \qquad (1.12)$$

Or, by rearranging Equation 12 algebraically,

$$R_T = \frac{1}{\dfrac{1}{R_1} + \dfrac{1}{R_2} + \dfrac{1}{R_3} + \ldots \dfrac{1}{R_N}} \qquad (1.13)$$

where N is the number of branches in the parallel circuit.

EXAMPLE 4

What is the total resistance of a parallel circuit with three branches: $R_1 = 50 \ \Omega$, $R_2 = 25 \ \Omega$, and $R_3 = 15 \ \Omega$?

Solution:
To calculate the total resistance, use the reciprocal equation:

$$R_T = \cfrac{1}{\cfrac{1}{50} + \cfrac{1}{25} + \cfrac{1}{15}} = 7.89 \ \Omega$$

Notice that the total resistance (R_T) is less than the smallest branch resistance (15 Ω).

Power calculations for parallel circuits are similar to the power calculations for series circuits. The total power in a parallel circuit is the sum of the power in the individual components of that circuit, just as the total power in a series circuit is equal to the sum of the power in the individual components of that circuit:

$$P_T = P_1 + P_2 + P_3 + \ldots P_N \tag{1.14}$$

Power calculations for these two types of circuits are also similar in that, in both circuits, the total power is the product of the source voltage and the total circuit current:

$$P_T = E_T \times I_T \tag{1.15}$$

1.8 Combination Circuits

The first step in analyzing a combination circuit is to reduce (simplify) the circuit as much as possible. Each section to be reduced will be a group of two or more resistors with the equivalent resistance results taking the place of the group. The series resistors should be reduced first. In the example circuit shown in Figure 1–6, only one series combination exists, and it must be reduced first. Subsequently, the parallel combinations are reduced.

Reducing a more complex circuit to its equivalent series resistor is performed much the same as discussed previously. The general approach is as follows:

1. Reduce only one part at a time
2. Ensure that all series resistors have been combined before a parallel portion is reduced
3. Combine parallel portions to a single resistor
4. Repeat combining equivalent resistors until all portions are reduced to one equivalent resistance

Once you know the total resistance in a combination circuit and the total source voltage, you can use Ohm's law to calculate the total current. However, the value of current through each respective resistor must be calculated in reverse order. Work backward from the total equivalent

FIGURE 1–6 Analyzing combination circuits.

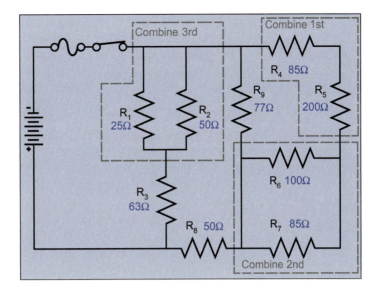

resistance calculated for Figure 1–6. As each equivalent branch resistance is found, apply the voltage to that branch and calculate the total branch current. Repeat the steps until all resistances, their voltage drops, and calculated currents are known.

As with other types of circuits we have studied, the total power utilized in a combination circuit (see Figure 1–6) is the sum of the power dissipated in each of the individual components in that circuit. It is also equal to the power used by the "equivalent resistance" of the circuit. If two of the three parameters of each component are known (voltage, current, and/or resistance), power can be calculated for each. Total power is then derived by adding all the individual powers:

$$P_{\mathrm{T}} = P_1 + P_2 + P_3 + \dots P_{\mathrm{N}} \tag{1.16}$$

where N is the number of components in the combination circuit.

■ WIRE CHARACTERISTICS

In the *DC Theory* text, you learned that the voltage drop E_{vd} across a load is equal to $I \times R$ and that the resistance of wire is found by using the formula

$$R = \frac{K \times L}{A_{\mathrm{cmil}}} \tag{1.17}$$

where R is the resistance of the wire in ohms, K is the resistivity of the wire in ohms per mil foot, L is the length of wire in feet, and A_{cmil} is the area of the wire in circular mils (cmil). This formula shows that wire resistance is fixed and does not exist as a condition or result of voltage or current. Another variation of the formula is used to calculate the voltage drop in a single-phase circuit. This variation is

$$E_{vd} = \frac{I \times K \times 2L}{A_{\mathrm{cmil}}} \tag{1.18}$$

Note that the multiplier of 2 takes into account that the wire makes a round-trip from the source to the load. Later you will learn to calculate the voltage drop of a three-phase circuit using a similar formula:

$$E_{vd} = \frac{I \times K \times \sqrt{3} \times L}{A_{cmil}}$$

(1.19)

The mysterious $\sqrt{3}$ is discussed later in this text.

■ MAGNETISM

There are three basic classifications of magnetic materials: *ferromagnetic, paramagnetic,* and *diamagnetic.* Both ferromagnetic and paramagnetic materials are metals. The magnetic properties of ferromagnetic materials allow the lines of magnetic flux to easily pass through. In ferromagnetic materials, such as iron and nickel, the lines of flux concentrate, or focus. The ability to easily concentrate these lines of flux makes the ferromagnetic materials the most easily magnetized. Paramagnetic materials, such as titanium, tend to somewhat block the passing through and concentration of the lines of flux and are therefore not magnetized as easily or as strongly. The resistance to the magnetic lines of flux is called *reluctance.* The best permanent magnets are made of steel, which is an alloy of iron and other metals. A magnet made with only soft iron does not hold its magnetism long.

The third type of material, diamagnetic material, can be made of metallic or nonmetallic materials. These materials do not allow any magnetic flux lines to pass through them but cause the magnetic lines of flux to pass around the diamagnetic material. A diamagnetic material that has a large value of reluctance is a good shield for magnetism. Examples of diamagnetic materials are brass and antimony.

1.9 Magnetism and Current Flow

When current flows through a wire, a magnetic field is produced around it in concentric circular form (see Figure 1–7). The two circles are cross-sectional views of wires, and both have current flowing

FIGURE 1–7   Magnetic field polarity of a conducting wire.

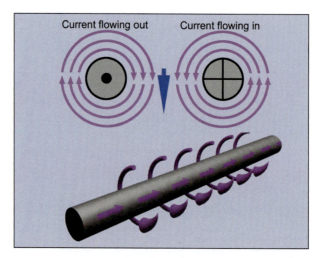

Current flowing out   Current flowing in

through them. The one on the left has current flowing "out of" the page. The right one has current flowing "into" the page. The arrows on the concentric circles show the polarity of the magnetic field created by the current. The field around the wire on the left has a clockwise polarity, while the one on the right is counterclockwise. The wire at the bottom shows the direction of the magnetic field with the current flowing from left to right. The direction of the field that encircles the conductor can be determined by using the left-hand rule for conductors. This rule states that if you wrap your left hand around the conductor, with your thumb pointing in the direction of current flow, your fingers will encircle the conductor in the direction of the magnetic field surrounding that conductor.

1.10 Magnetic Coils

If a wire is wound into a coil, as shown in Figure 1–8, and current is passed through it, the individual loops behave like small parallel wires having current in the same direction. This aiding current creates one large magnetic field, with a north pole and south pole, like a permanent magnet. Such coils are called *electromagnets.*
The intensity of the field depends on two factors:

1. The number of coils—a greater number of coils will create a larger magnetic field.
2. Current magnitude—a higher current creates a larger field around the wire and thus a larger overall field.

Multiplying these two factors results in ampere-turns, which represent the total number of magnetic lines created by the electromagnet.
The effect that an electromagnet has is only partially determined by its intensity. The density of the magnetism (usually expressed as number of lines per area) is a direct measure of how "strong" the electromagnet will be. In a coil, at least two factors are important:

1. Wrapping the coils closer together or reducing the coil radius will decrease the leakage flux and increase the density of the magnetic field. Figure 1–9 shows this effect.
2. If the coil were wrapped around a core of ferromagnetic material, the number of flux lines passing through the center of the coil would increase, and so would the strength of the electromagnet. This happens because the individual molecules in the core material become polarized (aligned) in relation to the magnetic field produced by the coil.

FIGURE 1–8 Magnetic field produced by current in a coil.

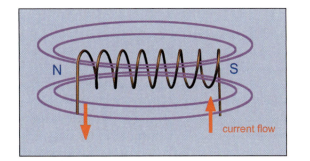

current flow

FIGURE 1–9   Increasing field strength of closer (or smaller) coils.

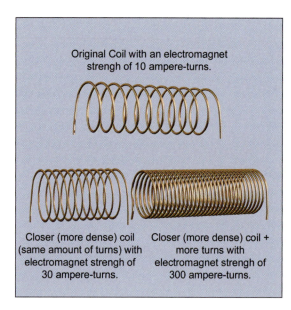

Original Coil with an electromagnet strengh of 10 ampere-turns.

Closer (more dense) coil (same amount of turns) with electromagnet strengh of 30 ampere-turns.

Closer (more dense) coil + more turns with electromagnet strengh of 300 ampere-turns.

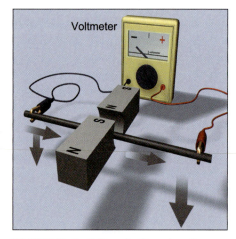

Voltmeter

FIGURE 1–10   Induced voltage causing current flow.

1.11 Inducing Voltage with Magnetism

A magnetic field is generated around a current-carrying conductor. A reverse principle states that a voltage will be induced into a conductor that is passed through a magnetic field. If this conductor is connected to a complete path, current will flow.

Look at Figure 1–10. The conductor is being moved downward between the poles of the two magnets. The poles of the magnets create a field that is moving from north to south. This downward motion of the conductor causes a voltage to be induced into the conductor, the polarity of which causes the current to flow in the direction shown. Figure 1–11 shows the so-called left-hand rule that can be used to determine the polarity of current flow induced in a wire when it cuts or is cut by a magnetic field. The amount of voltage being induced into a wire is determined by three main factors:

1. The strength of the magnetic field
2. The speed of the wire cutting the magnetic field
3. The number of loops in the wire

1.12 AC and DC Generators

Look at Figure 1–12. This simple **alternating current (AC)** generator is constructed of a permanent magnet, slip rings, brushes, and a single loop (armature). The armature is the rotating part of the machine in which the voltage is induced. The magnets set up the required magnetic field. As the loop rotates, each side of the loop will be cutting lines of flux at the same time. As the lines of flux are cut, a voltage is produced at the ends of the loop. The voltage is connected to the voltmeter by the slip rings. A slip ring is attached to each end of the loop.

DC generators work basically the same as AC, except the voltage is removed from the armature by a commutator instead of slip rings. Look

FIGURE 1–11   The left-hand rule for generators.

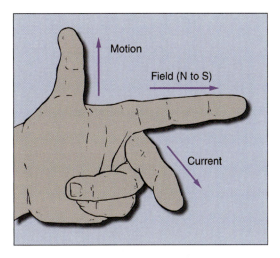

FIGURE 1–12   AC voltage generator.

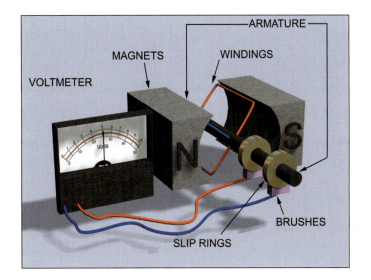

at Figure 1–13. Each commutator segment is attached to the end of one side of each loop in the generator. As the commutator segment aligns with each brush, the current flow is always in the same direction. The left brush in Figure 1–13 passes the current from the left side of the loop, and the right brush passes the current from the right side of the loop. This means that the left brush is always negative and the right always positive, even though the loops change polarity as they rotate.

■ THEOREMS AND LAWS

A variety of theorems are available to help you in your analysis of electrical circuits. Each of these theorems was covered in detail in your *DC Theory* text. They are mentioned here to encourage you to review them if you have forgotten their application and use. Each of these theorems works just as well in AC circuits as in DC.

FIGURE 1–13   DC voltage generator.

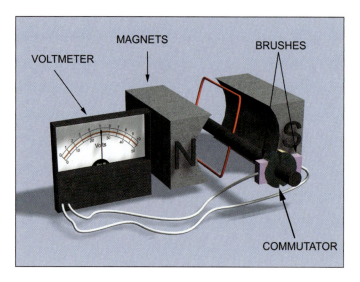

1.13 Superposition Theorem

The superposition theorem states that the voltage or current in any element resulting from several sources acting together is the sum of the voltages or currents resulting from each source acting alone. To calculate the effects from a single source, all other sources must be turned off while the first source's effects are calculated. This means disabling the other sources so that they cannot generate voltage or create current. This has to be done without changing the resistance of the circuit. You can do this by assuming that the voltage sources not being measured are shorted across their terminals and that the current sources are open circuited.

The steps in applying the superposition theorem to a two-voltage source circuit are the following:

1. Reduce all but one source (voltage or current) to zero by replacing the voltage sources with a short and the current sources with an open
2. Calculate the total resistance, voltage drops, and current for the redrawn circuit, noting the polarity of the voltage and the direction of the current flow
3. Repeat steps 1 and 2 for all the voltage and current sources in the circuit
4. Calculate the algebraic sum of the currents and voltage drops in each circuit element by adding each individual component algebraically

1.14 Kirchhoff's Laws

Kirchhoff's laws provide you with more tools to solve more complex circuits than is possible using Ohm's law. Applying these laws allows you to gain a better working knowledge of circuit operations necessary for solving multiple-source circuits. Kirchhoff's laws are *not* replacements for Ohm's law; rather, they augment Ohm's law by going beyond the basics and giving you new skills in circuit analysis. Kirchhoff's two laws are the following:

1. The algebraic sum of the voltages around any closed path is zero.
2. The algebraic sum of the currents entering any node (junction point) is zero.

In applying Kirchhoff's voltage law, the first step is to determine the algebraic signs for the voltages. One way of doing this is to go around one of the closed loops and treat any voltage whose positive terminal is reached first as positive and any negative terminal reached first as negative. This applies to the voltage drops and voltage sources. Label all the terminals until you return to the starting point. The algebraic sum of all the voltage terms must be zero.

Loop equations are written so that the algebraic sum of the *IR* voltage drops equals zero. Care must be taken to observe the polarity and direction of current flow. Remember that the point where current enters a resistive element is considered negative.

1.15 Thevenin's Theorem

Thevenin's theorem is used to reduce any circuit, as viewed from any two terminals in that circuit, to an equivalent voltage source in series with a resistor. This theorem is illustrated in Figure 1–14. The complex circuit network can all be represented by an equivalent series circuit with respect to any "pair" of terminals in the network.

Look at Figure 1–14. Imagine that the block on the left contains a network connected to terminals a and b. Thevenin's theorem states that the entire network connected to a and b can be replaced by a single voltage source (V_{TH}) in series with a single resistance (R_{TH}) connected to the same terminals. The Thevenin circuit assumes the output voltage to be the open-circuit voltage with no "load" resistance connected. To calculate the equivalent Thevenin's resistance, assume that the voltage source is shorted (removed) and measure resistance across the series resistor (R_{TH}). Although the source (V_{TH}) is shown as a battery, Thevenin's theorem will work equally well for an AC circuit.

1.16 Norton's Theorem

This theorem is similar to Thevenin's theorem and differs in only two ways:

1. Norton's theorem uses a current source.
2. The equivalent resistance is placed in parallel with the source instead of in series.

FIGURE 1–14 Thevenin's theorem.

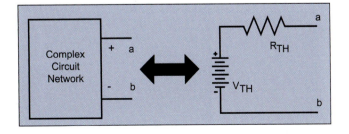

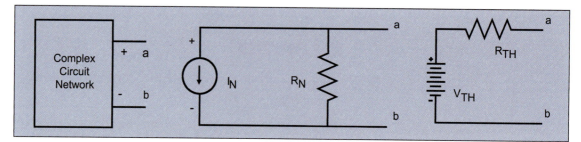

FIGURE 1–15   Comparison between Norton's theorem and Thevenin's theorem.

Figure 1–15 shows the comparison between Thevenin and Norton circuits.

The Norton's current source is determined shorting the a and b terminals. The Norton resistance is calculated in the same way as the Thevenin resistance.

■ SUMMARY

The basis of all electrical theory is found in the atom. The key electrical parts of an atom are the proton (positive charge) and the electron (negative charge). The electron plays the central role in the generation transmission, distribution, and usage of electricity.

All materials fall into one of three categories: conductors, semiconductors, and insulators. Conductors are used to carry electrical energy from the source of generation to the load that needs it. Conductors are usually made of a metallic substance, such as copper or aluminum. The ability of a conductor to carry electrical current is determined by its resistivity and its cross-sectional area. This is described by the formula

$$R = \frac{K \times L}{A} \qquad (1.20)$$

After the resistances and the sources are known, voltages and currents can be determined in any part of the circuit by careful application of a variety of laws and theorems. Ohm's law is at the heart of all these methods. Each of them, however, offers its own special use in different circuit situations.

■ REVIEW QUESTIONS

1. Discuss the relationship among the electron, the proton, and electricity.
2. How is the resistance of a conductor related to its diameter, its cross-sectional area, and its resistivity?
3. Discuss how current, voltage, and current are arranged in series, parallel, and combination circuits.
4. What is the relationship between current flow and magnetism?
   a. Polarity
   b. Magnitude
5. How can the magnetic field of a coil be strengthened?

6. Discuss the left-hand rule for generators. How can it be used to determine the polarity of a generator voltage output?
7. What is the principal difference between an AC generator and a DC generator?
8. Discuss the following theorems and explain how each is used in the analysis of an electrical circuit:
   a. Superposition theorem
   b. Kirchhoff's laws
   c. Thevenin's theorem
   d. Norton's theorem

■ PRACTICE PROBLEM

1. Look at Figure 1–6 and assume that the power supply is a 120-V-battery. Calculate the following unknowns:

   a. Total circuit current

   b. Total circuit power

   c. Voltage drop, current flow, and power dissipation for each of the nine resistors

chapter **2**

Mathematics: Using Vectors Effectively

■ OUTLINE

■ OVERVIEW

Circuit analysis and calculations with resistors can be easily performed using simple arithmetic addition, subtraction, multiplication, and division. The only concern is to be certain that you pay attention to the sign of the calculation, that is, negative or positive. This is true whether the circuit being analyzed is AC or DC.

Mixing AC, resistors, inductors, and capacitors into your calculations complicates the issue. Analysis of such circuits requires a more sophisticated approach to the mathematics, namely, the use of vectors.

In this chapter, you will learn how to solve systems vectors by adding, subtracting, multiplying, and dividing them. A proper understanding of vectors is necessary for you to understand the behavior of the variables in AC circuits that contain resistance, inductance, and capacitance.

Look at Figure 2–1. The **reactance** (3 Ω) and **resistance** (4 Ω) in a circuit are at a 90° angle to each other and are acting on a common point. Vector analysis provides the means for calculating the combined or net effect of these two forces. In this example, the resulting combination is equal to a magnitude of 5 Ω and an angle of 36.9° from the horizontal axis. Vectors are also used to show the phase relationships between voltage and current in AC circuits; therefore, you must rely heavily on vector mathematics to solve circuits having inductive and capacitive components.

This chapter also covers information on solving for the resultant vector value for any combination of multiple vectors using several different methods to solve those problems.

■ OBJECTIVES

After completing this chapter, you should be able to:

1. Describe the differences and similarities among scalars, vectors, and phasors.
2. Add, subtract, multiply, and divide vectors and phasors.
3. Explain the relationships among phasors, vectors, and complex numbers.

FIGURE 2–1 Vector result of a reactance and a resistance in the same circuit.

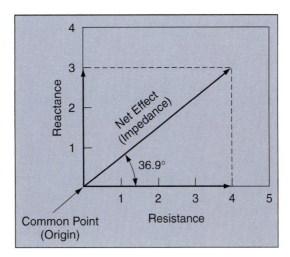

■ GLOSSARY

Hypotenuse   The side of a right triangle that is opposite the right angle.

Phasor   A vector that rotates. Phasors are used to describe voltages, currents, and other such quantities in electrical systems.

Polar form   Representation of a vector as a magnitude and an angle.

Reactance   The amount of opposition to current flow exhibited by a magnetic field in an inductor or an electrostatic field in a capacitor.

Rectangular form   Representing a vector as the sum of two other vectors that are at right angles to each other.

Resistance   The amount of opposition to current flow through a material caused by so-called frictional effects.

Scalar   A number with magnitude only. For example, 5 miles per hour is a scalar.

Vector   A number with both magnitude and direction. For example, 5 miles per hour going due east is a vector.

FIGURE 2–2   Scalar path sign.

FIGURE 2–3   Vector path sign.

■ SCALARS, VECTORS, AND PHASORS

2.1 Scalars

Scalars are numbers used to represent magnitude only and do not take direction into consideration. Assume that you are lost in the woods. As you walk down the path, you come upon a sign like that shown in Figure 2–2. Does this sign help? It tells you that it is 3 miles to the highway, but in which direction? This sign would represent a scalar because it contains only a magnitude. If the sign were firmly planted in the ground and had an arrow painted on it pointing toward the highway, it would become a vector because it contains both the magnitude and the direction. Figure 2–3 represents a vector having both magnitude (3 miles) and direction (direction of arrow).

2.2 Vectors

A **vector** is a symbol that indicates both magnitude and direction. A vector tells not only "how much" but also "in what direction." A vector is represented graphically by an arrow. The length of the arrow represents the magnitude. The tip of the arrow represents the direction of the vector and is identified by its angle of rotation from 0°. For example, the direction of the vector in Figure 2–1 is 36.9° counterclockwise from the horizontal axis.

2.3 Phasors

A **phasor** is a vector that rotates. A voltage and/or a current in an AC circuit will be represented by a phasor that rotates at the frequency of the AC waveform. Although phasors are somewhat more powerful and used differently than vectors, manipulation of them in the circuit analysis that you will be working with is identical to vectors. The remainder of this chapter and text refers to all electrical quantities as vectors even though some of them are, strictly speaking, phasors.

2.4 Vector Referencing

As stated previously, a vector is drawn as a straight line with an arrow placed at one end. The arrow represents the direction of the vector, and the line length represents the magnitude of the vector. The vector can represent any quantity, such as inches, miles, volts, amps, ohms, or power.

The zero reference (0°) is a horizontal line to the right. The direction or angle of other vectors is positive when measured in a counterclockwise direction (see Figure 2–4). In Figure 2–4, a vector with a magnitude of 4 is at angle 0°. The second vector is drawn with a magnitude of 3 at an angle of 45° from the first vector. Now the third vector is referenced from the first vector at an angle of 90°. It is important to note that the third vector is referenced from the 0° line and not the previous vector at 45°.

Notice that the three vectors could also be referenced in a clockwise direction using the negative values. The first vector would still be 4 at 0°, the second would be 3 at −315°, and the third would be 4 at −270°.

FIGURE 2–4 Vector counterclockwise rotation.

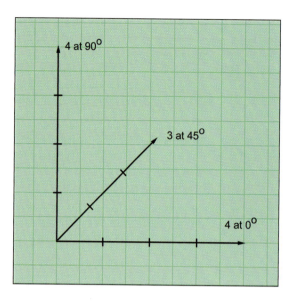

2.5 Rectangular and Polar Representation

So far in this chapter, all the vectors have been represented in what is called **polar form**: a magnitude and an angle. Vectors used in AC circuit analysis can also be referenced by expressing them as the vector sum of two other vectors that are at 90° from each other. This is called **rectangular form**.

For example, the vector shown in Figure 2–1 can be expressed in polar form as 5 at an angle of 36.9°. This is usually written as $5\angle\ 36.9°$. The vector in Figure 2–1 can also be represented as the sum of one vector at 0° plus another vector at 90°. In this case, the 0° vector is 4 Ω long, and the 90° vector is 3 Ω long. The entire expression is usually given as $4 + i3$ or $4 + j3$, where $+i$ or $+j$ means that the second vector goes up vertically or 90°.

■ RIGHT TRIANGLES AND TRIGONOMETRY

To manipulate vectors effectively, the electrician must be familiar with the characteristics and behavior of right triangles and the relationships of the sides of a right triangle. The next two sections describe these important concepts and will serve as a refresher for your (probably) long-forgotten trigonometry studies.

2.6 Right Triangles

Look at Figure 2–5. A right triangle is a triangle that has a 90° angle. The **hypotenuse** is the longest side and is always the side that is opposite the 90° angle. Pythagoras of Samos figured out that the sum of the square areas bordered by each of the two sides of the right triangle would be equal to the square area bordered by the hypotenuse, or longest side. This is stated mathematically as

$$C^2 = A^2 + B^2 \tag{2.1}$$

FIGURE 2–5 Right triangle with the sides squared.

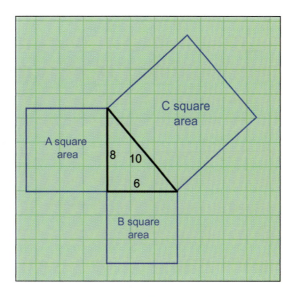

Where:

   C = the length of the hypotenuse
   A = the length of one side
   B = the length of the other side.

Equation 2.1 is called the Pythagorean theorem in honor of its discoverer.

Using the Pythagorean theorem, the hypotenuse in Figure 2–5 can be found as

$$8^2 + 6^2 = C^2 \tag{2.2}$$

$$64 + 36 = C^2 = 100 \tag{2.3}$$

$$C = 10 \tag{2.4}$$

EXAMPLE 1

Look at Figure 2–6. Which side is the hypotenuse? What is the length of the hypotenuse if two of the sides have a value of 3 and 6?

Solution:

$$C^2 = A^2 + B^2$$
$$C^2 = (6)^2 + (3)^2$$
$$C^2 = 36 + 9$$
$$C^2 = 45$$
$$C = \sqrt{45} = 6.71$$

2.7 Sines, Cosines, and Tangents

Since the sides of a right triangle have a mathematical relationship to each other (the Pythagorean theorem), it should come as no surprise that the ratios of the various sides of the right triangle also have very specific values.

FIGURE 2–6 Example of Pythagorean theorem.

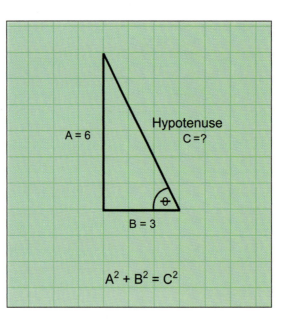

Figure 2–7 is a right triangle with the various sides and angles labeled. Each angle has a side opposite and a side adjacent to it. Table 2–1 shows the relationships of each angle to each of the sides. Use Figure 2–7 and Table 2–1 to understand the definitions given in Table 2–2.

Several memory tricks are often used to remember this relationship. **O**scar **H**ad **A H**eap **O**f **A**pples or **O**h **H**eck **A**nother **H**our **O**f **A**gony are two of the many. They work as shown in Equations 2.5 to 2.7:

$$\frac{\text{Opposite (Oscar)}}{\text{Hypotenuse (Had)}} = \sin \qquad (2.5)$$

$$\frac{\text{Adjacent (A)}}{\text{Hypotenuse (Heap)}} = \cos \qquad (2.6)$$

FIGURE 2–7   Sample triangle for sine, cosine, and tangent.

Table 2–1   Angles and Their Related Sides (see Figure 2–7)

Angle	Side Opposite	Side Adjacent
A	a	b
B	b	a

Table 2–2   Table of Trigonometric Definitions (see Figure 2–7)

Angle	Sine	Cosine	Tangent
A	$\sin(A) = \dfrac{\text{Opposite}}{\text{Hypotenuse}} = \dfrac{a}{c}$	$\cos(A) = \dfrac{\text{Adjacent}}{\text{Hypotenuse}} = \dfrac{b}{c}$	$\tan(A) = \dfrac{\text{Opposite}}{\text{Adjacent}} = \dfrac{a}{b}$
B	$\sin(B) = \dfrac{\text{Opposite}}{\text{Hypotenuse}} = \dfrac{b}{c}$	$\cos(B) = \dfrac{\text{Adjacent}}{\text{Hypotenuse}} = \dfrac{a}{c}$	$\tan(B) = \dfrac{\text{Opposite}}{\text{Adjacent}} = \dfrac{b}{a}$

Note: Sine is abbreviated as "sin," cosine is "cos," and tangent is "tan."

$$\frac{\text{Opposite (Of)}}{\text{Adjacent (Apples)}} = \tan \qquad (2.7)$$

Some believe that remembering these clever memory devices is more difficult than remembering the original definitions. You should use whatever method is easiest for you.

After the sine, cosine, or tangent of the angle is known, the angle itself can be found using the trigonometric functions on a scientific calculator or by the trigonometric tables.

EXAMPLE 2

The hypotenuse of the triangle in Figure 2–8 is 14 and side *a* is 9. How many degrees are in ∠ *A*?

Solution:
Since the lengths of the hypotenuse and the opposite side are known, the sine function can be used:

$$\sin(A) = \frac{\text{Opposite}}{\text{Hypotenuse}} = \frac{9}{14} = 0.643$$

But 0.643 is not the angle; it is the sine of the angle.

You must use a scientific calculator and the inverse sine function (sometimes called the arcsine) to determine the angle. The way you determine the inverse sine depends on the type of calculator you are using. Either enter 0.643 and press the arc or inv sine button or enter arc sine(0.643) and press the equals (=) sign. The answer is arcsine(0.643) = 40°.

The same process is used when utilizing the cosine function or the tangent function.

FIGURE 2–8   Triangle for Example 2.

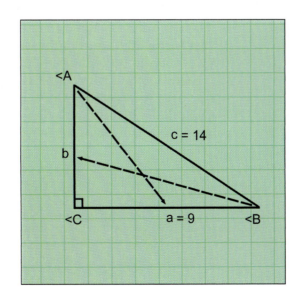

EXAMPLE 3

Using the same triangle in Figure 2–8, find the number of degrees in ∠ B.

Solution:

$$\cos(B) = \frac{\text{Adjacent}}{\text{Hypotenuse}} = \frac{9}{14} = 0.643$$

The scientific calculator shows that the arccosine of 0.643 is 50°.

Note in this problem that the opposite side became the adjacent side to the hypotenuse because the reference angle changed. For this problem, side a is the base of the triangle. Also note that the decimal (0.643) remained the same because of the ratio (9/14), but the angle changed because of the use of the cos function instead of the sin function. One last point should be noted. All the angles in a right triangle must total 180°. This means that the right angle (∠C = 90°) + ∠A + ∠B = 180°. If you knew from the first problem that ∠A was 40°, then you could calculate the remaining angle by simple addition and subtraction—since 90 + 40 = 130, and 180 − 130 = 50, then ∠B = 50°.

■ VECTOR ADDITION AND SUBTRACTION

Vectors can be added and subtracted in a variety of arithmetic and geometric ways. The following sections describe those methods and give examples of how each may be employed.

2.8 Vectors in the Same Direction

Since vectors can be used to represent quantities such as volts, amps, ohms, and power, they can be added, subtracted, multiplied, and divided. There are several methods to perform vector addition. Regardless of the method used, they must be added with a combination of geometric and algebraic addition. This is called *vector addition*. (Note that vector addition is quite easy on most scientific calculators; however, you should understand the fundamentals.)

One method is to connect one vector to the endpoint of the other one. This works easiest when the vectors are in the same direction. (see Figure 2–9). In this figure, three vectors with the same angle (0°) are being added. The total sum is 7 + 5 + 3 = 15.

FIGURE 2–9   Simple vector addition.

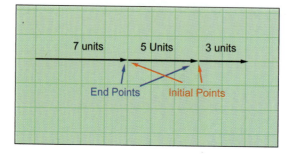

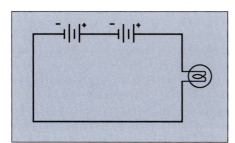

FIGURE 2–10   A flashlight with two batteries.

Now look at Figure 2–10 and consider having two batteries connected in series the way they are in a flashlight. The bulb in the flashlight is designed to operate on 3 volts. Since the standard flashlight cell is only 1.5 volts, two batteries must be added together to make 3 volts. In vector terms, it would look like one vector of 1.5 volts plus another vector of 1.5 volts (see Figure 2–11).

2.9 Vectors in Opposite Directions

To add the vectors that are in opposite directions (180° apart), subtract the magnitude of the smaller from the magnitude of the larger. The result is a vector in the same direction as the vector with the larger magnitude. Assume that the batteries in Figure 2–10 are 3 volts and 5 volts, respectively. Somehow one of the batteries was placed in the flashlight backward. The voltages would oppose each other. This means that 3 volts of the 5-volt battery would try to overcome the 3 volts of the other battery. The result would be a vector with a magnitude of 2 volts in the same direction as the 5-volt battery (see Figure 2–12).

In numerical form, this equation would look like adding a positive number to a negative number:

$$+5 + (-3) = +2 \qquad (2.8)$$

Since adding a negative is the same as subtracting, Equation 2.15 reduces to

$$5 - 3 = 2 \qquad (2.9)$$

2.10 Vectors in Different Directions

Vectors having directions other than 0° and 180° from each other can also be added. An example is shown in Figure 2–13. A vector with a magnitude of 4 and a direction of 20° and a vector with a magnitude of

FIGURE 2–11   Two 1.5-volt vectors added together to get a total of 3 volts.

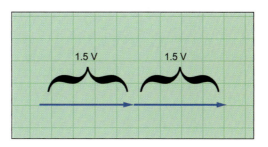

FIGURE 2–12   Adding vectors with opposite directions (180°).

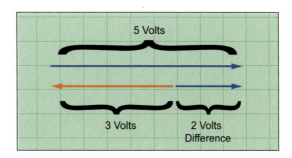

FIGURE 2–13   Adding vectors with different directions.

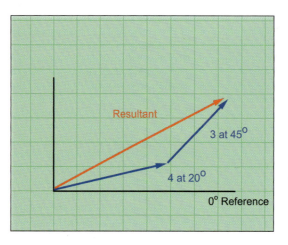

3 and a direction of 45° are added together. The addition is made by connecting the starting point of the second vector to the endpoint of the first vector. The direction of the vector, stated as the number of degrees of the angle, is always referenced to the x-axis which will always be 0°. The resultant is drawn from the starting point of the first vector to the endpoint of the second. This is called the *triangular method*. It is possible to add more than two vectors in this fashion.

2.11 Parallelogram Method

The parallelogram method can be used to find the resultant of two vectors that start at the same point instead of connecting the starting point of the second vector to the endpoint of the first vector. A parallelogram is a four-sided figure whose opposite sides are parallel to each other. A rectangle is a parallelogram with 90° angles.

For example, consider the vectors $10\angle25°$ and $12\angle55°$. Remember that the vectors must begin at the same point and that all angles are referenced off the x-axis (0°). To find the resultant of these two vectors, form a parallelogram using the vectors as two of the sides (see Figure 2–14). The resultant is drawn from the corner of the parallelogram where the two vectors intersect to the opposite corner.

FIGURE 2–14   The parallelogram method.

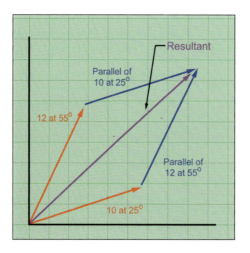

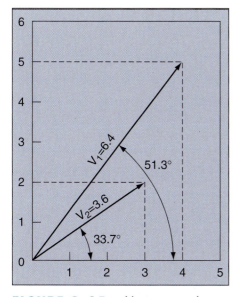

FIGURE 2–15   Vector notation.

2.12 Using the Rectangular Components

As was hinted at previously, vectors can be represented as the sum of their two components (called rectangular form or rectangular notation) or as a magnitude at an angle (polar notation). Figure 2–15 shows two vectors.

Table 2–3 shows the two ways that each vector can be represented. Be aware that the letter j (electricians use j instead of i) is itself a vector defined as $j = 1\angle 90°$.

EXAMPLE 4

Add the vectors V_1 and V_2 in Figure 2–15.

Solution:
First the problem is solved by entering the polar form into a scientific calculator and adding the results (the Hewlett Packard model HP48GX was used for this example):

$$V_1 + V_2 = V_t = 6.4\angle 51.3° + 3.6\angle 33.7° = 9.89\angle 45°$$

Next, add the vectors using their rectangular components. The sum of the two vectors is equal to the sum of their horizontal components added to the sum of their vertical components:

$$V_1 + V_2 = V_t = (4 + j5) + (3 + j2) = (7 + j7)$$

Are the two answers the same? For the answer, refer to Figure 2–16. This figure is the vector (V_T) as determined by adding the horizontal and vertical parts of the two component vectors. Notice that V_T forms a right triangle with the horizontal and vertical axes. This means that the tan of the unknown angle (??) is equal to 7/7. In other words, ?? = arctan(7/7) = 45°. What about the length? From the Pythagorean theorem, $\rightarrow V_t = \sqrt{7^2 + 7^2} = \sqrt{98} = 9.89$.

■ MULTIPLYING AND DIVIDING VECTORS

You will be called on to multiply and divide vectors in your work as an electrician. For example, the use of Ohm's law requires vectors for voltage, current, and impedance (AC resistance). To multiply two vectors, you multiply their magnitudes and add their angles. To divide two vectors, you divide their magnitudes and subtract their angles.

Table 2–3   Polar and Rectangular Formats for Vectors of Figure 2–15

Vector	Polar	Rectangular
V_1	$6.4\angle 51.3°$	$4 + j5$
V_2	$3.6\angle 33.7°$	$3 + j2$

FIGURE 2–16 Sum of vectors in Figure 2–15.

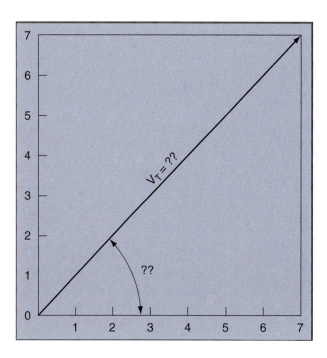

EXAMPLE 5

Given: $V_1 = 5\angle 20°$ and $V_2 = 4\angle 30°$

Problem 1: What is the product of V_1 times V_2?

Solution:

$$V_1 \times V_2 = 5\angle 20 \times 4\angle 30 = (5 \times 4)\angle 20 + 30 = 20\angle 50$$

Problem 2: What is the quotient of V_1 divided by V_2?

Solution:

$$\frac{V_1}{V_2} = \frac{5\angle 20}{4\angle 30} = \left(\frac{5}{4}\right)\angle 20 - 30 = 1.25\angle -10°$$

■ SUMMARY

Scalar quantities have only magnitude. Vectors are lines that indicate both direction and magnitude. Vectors can be added, subtracted, multiplied, or divided. The direction of a vector is indicated by the arrowhead at the end. When a single vector is produced from combining two or more vectors, it is called a *resultant*. From zero degrees, vectors rotate in a counterclockwise direction.

The sum of all the angles in a right triangle is 180°. The relationship between the length of the sides of a right triangle to the number of degrees in its angles can be expressed as the sine, cosine, or tangent of a particular angle.

The sine function is the relationship of the opposite side divided by the hypotenuse. The cosine function is the relationship of the adjacent side divided by the hypotenuse. The tangent function is the relationship of the opposite side divided by the adjacent side.

The hypotenuse is always the longest side of a right triangle. The opposite and adjacent sides are determined by which angle is being used as the reference angle.

■ REVIEW QUESTIONS

1. On a sheet of paper, list the similarities and differences among scalars, vectors, and phasors.
2. Describe the various methods for adding or subtracting vectors, including the following:
   a. Vectors in the same direction
   b. Vectors in the opposite direction
   c. Vectors in a different direction
   d. Adding by the rectangular components

3. Describe how to multiply and divide vectors.
4. Define, in your own words, the following terms:
   a. Sine
   b. Cosine
   c. Tangent
   d. Hypotenuse
   e. Side opposite
   f. Side adjacent

■ PRACTICE PROBLEMS

1. Add, subtract, multiply, and divide the following vectors:
   a. $(5 + j4)$, $(3 - j2)$
   b. $3\angle 35°$, $\left(2 - j\dfrac{2}{3}\right)$
   c. $15\angle -124°$, $32\angle 15°$
   d. $(3 - j4)$, $(3 + j4)$
   e. $(10 + j0.5)$, $25\angle 75°$

2. A vector has both _____ and direction.

3. Draw V_1 and V_2, where $V_1 = 20\angle 0°$ and $V_2 = 30\angle 30°$.

4. Show the sum of $V_1 + V_2$ using the parallelogram method.

5. Find angles A and B in the following figure:

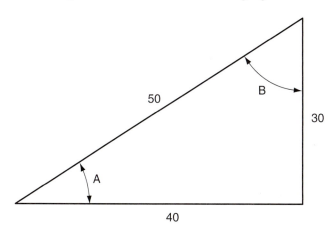

6. Answer the following questions about the following graph:
   a. What are the X and Y coordinates of the dot?
   b. If a line is drawn from the zero axis and the dot, what is the length of the line?

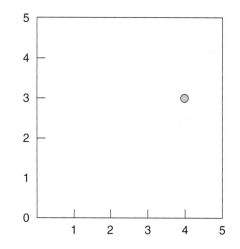

chapter 3

Comparing AC to DC

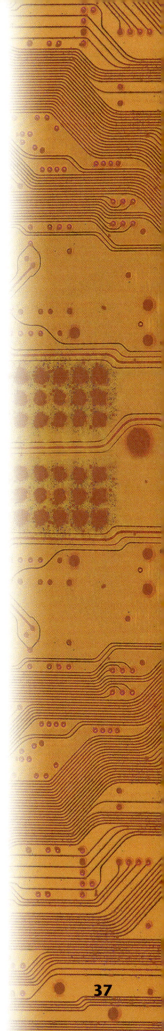

■ OVERVIEW

In the process of studying DC, you learned that the direction of current flow is always in the same direction—a smooth, continuous flow of electrons. Alternating current changes direction periodically. To better understand AC and DC, consider Figure 3–1.

Curve 1 is pure DC. Note that the waveform is a straight line stretching from left to right. It is at 2.5 V and might represent the voltage as measured on a 2.5-V battery.

Curve 2 is pure AC. Notice that the voltage magnitude spends exactly the same amount of time above the zero axis as it does below. This means that the electrons spend the same amount of time traveling in one direction as they do the other. The voltage is alternating.

Curve 3 is different. Note that it spends more time above the zero axis than it does below. It has an average (DC) value of approximately 0.75 V. Since is does change, it cannot be called pure DC; however, since it has a nonzero average value, it cannot be called pure AC. In fact, curve 3 is a mixture of AC and DC.

This chapter addresses pure AC waveforms and compares them with pure DC waveforms.

■ OBJECTIVES

After completing this chapter, you should be able to:

1. Define the terms cycle, frequency, period, alternation, sine wave, wavelength, instantaneous values, effective value, average value, maximum value, and peak-to-peak value.
2. Mathematically calculate values of various AC parameters.

FIGURE 3–1 AC and DC waveforms.

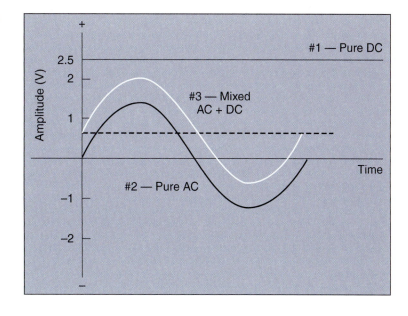

■ GLOSSARY

Apparent power   The total rate at which energy in a system is being used and borrowed. Mathematically, the apparent power is equal to the voltage times the current.

Average value   The mathematical average of the positive and negative values of an AC waveform. The value read by a DC responding meter. A pure sine wave has an average value of zero.

Eddy currents   Local current flow (usually undesirable) that is set up within a conductor because of stray magnetic fields. Eddy currents cause losses in the conductor.

Frequency   The rate at which an AC waveform cycles through its complete alternation. Measured in cycles per second (hertz).

Peak value   The maximum magnitude (positive or negative) of an alternating waveform.

Peak-to-peak value   The maximum distance from the negative peak to the positive peak of an AC waveform. For a sine wave, the peak-to-peak value is two times the peak value.

Period   The time required for an AC waveform to complete one full alternation.

Phase angle   The number of electrical degrees between equivalent points on two waveforms.

Reactive power   The power in an electrical circuit that is used temporarily to establish magnetic or electrostatic fields. Reactive power represents energy that is borrowed by the inductors and capacitors in the system and then returned to the source.

RMS value   Root-mean-square value. The DC magnitude that would be required to accomplish the same amount of work as an AC waveform. For a sine wave, $RMS = Peak/\sqrt{2}$.

Skin effect   The tendency of alternating current to flow toward the outside, or "skin," of a conductor. Skin causes a decrease in the effective cross-sectional area available for current flow.

True power   The power in an electrical circuit that does real work, such as heat, light, or motion. Real power is the rate at which an electrical circuit or system is expending energy.

Volt-amperes   The unit of measure for apparent power. Abbreviation is VA.

Volt-amperes reactive   The unit of measure for reactive power. Abbreviation is VAR.

Watt   The unit of measure for real power. Abbreviation is W.

■ WHY ALTERNATING CURRENT?

The biggest single advantage of AC is that it can be transformed from one voltage to another; that is, a transformer can step voltage up or down. Higher voltages are better for transmitting electricity over long distances. This is because the higher voltage can transmit the same amount of power at a lower current value (remember that $P = EI$). Consequently, there is less heat loss (I^2R), and smaller, less expensive wire may be used. Higher voltages would be dangerous inside the home or the workplace, so transformers are used to step the voltage down to lower, safer levels. DC also has some advantages; however, AC is used for the overwhelming majority of major electrical energy supply systems.

■ SINE WAVES

The shape of the waveform of the voltage produced by a simple AC generator plays an important part throughout AC circuit theory. The AC generator's production of a sine wave is a direct result of the wire loop cutting through the lines of flux throughout its complete 360° rotation. The changes in direction take place as the generated voltage output, which starts at zero, increases to a maximum value, decreases and crosses zero from positive to negative, and then returns to zero all within that one rotation (see Figure 3–2). The name *sine wave* is based on the fact that the value of the voltage at any instant in time (V_I, the instantaneous value) within the rotation can be calculated by taking the peak voltage times the sine of the angle of travel of the loop at that specific instant: $V_I = V_{peak} \times \sin(\theta)$.

The variations in the voltage depend on the angle of the loop relative to the direction of the lines of flux. When the loop is moving perpendicular to the lines of flux, it is cutting the most lines of flux per second, and the maximum voltage is being generated. When the loop is moving parallel to the lines of flux, it is cutting zero lines of flux per second, and the voltage generated is zero. As the loop begins to travel

FIGURE 3–2   A complete sine wave.

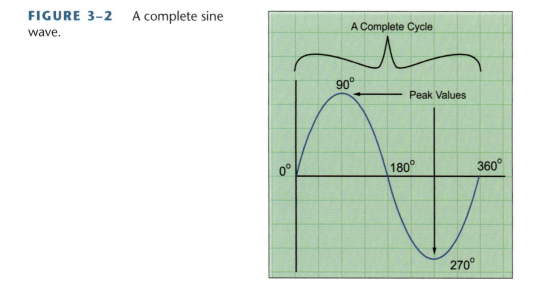

FIGURE 3–3   Complete sine wave cycle output from a generator.

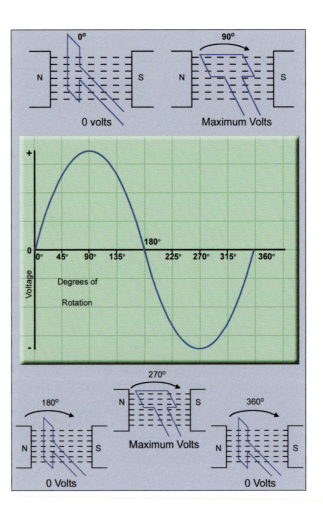

from a position parallel to the lines of flux toward a perpendicular position relative to the flux lines, an increasing number of lines of flux are cut each second, and the voltage increases (see Figure 3–3).

3.1 Frequency

As seen in Figure 3–2, each complete revolution of the loop through its 360° travel (one revolution) is called a *cycle*. Frequency, measured in hertz (Hz), is a measure of the number of cycles per second. If the voltage generated completes three complete cycles in 1 second, the frequency of that voltage is said to be 3 Hz (see Figure 3–4).

FIGURE 3–4   A 3-Hz sine wave.

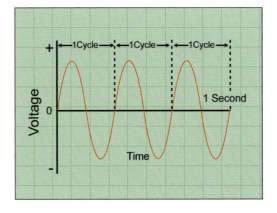

The most common frequency in North America is 60 Hz. This means that the voltage starts at zero and rises to its maximum value in the positive direction, returns to zero, travels to its maximum value in the negative direction, and then returns to zero 60 times in 1 second. Outside North America, the most commonly used frequency is 50 Hz.

3.2 Sine Wave Values

There are four forms of measuring the magnitude of voltage and current sine waves:

1. Zero to peak (sometimes just called "peak")
2. Peak to peak
3. RMS
4. Average

The **peak value** is measured from zero to the highest value obtained in either the positive or the negative direction. Therefore, the **peak-to-peak value** is twice that of the peak because it considers both positive and negative peaks (see Figure 3–5).

The **RMS value** is also called the *effective value* because it compares the amount of energy available from an AC waveform to DC waveform. Since the AC waveform is at its peak for only an instant and then returns through its cycle, using the peak value would not provide an accurate method of measurement. The RMS value can be used to compare the power output or heating value of the AC waveform voltage. Put another way, when applied to a light, what level of AC current will produce the same amount of light as a 5-ampere DC current? "RMS" stands for "root mean square," which is a shortened way of saying "square root of the mean (average) of the square" of the instantaneous currents. For a sine wave, the RMS value can be found by dividing the peak by the square root of 2 (1.414) or by multiplying by the inverse of 1.414 (0.707).

FIGURE 3–5   Peak and RMS values of a sine wave.

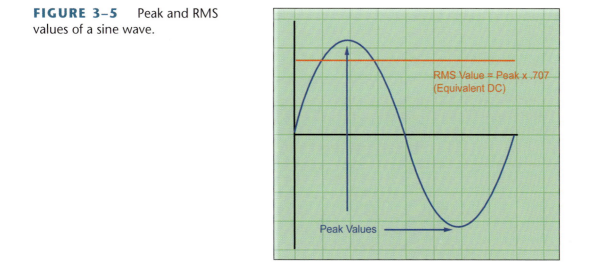

RMS Value = Peak x .707
(Equivalent DC)

Peak Values

EXAMPLE 1

An AC waveform has a peak value of 423 amps. What value of DC current would produce the same amount of heat in the wires carrying the current?

Solution:
The effective value of the AC waveform is the peak value divided by the square root of 2:

$$I_{\text{RMS}} = \frac{I_{\text{peak}}}{\sqrt{2}} = 299\ A$$

EXAMPLE 2

The RMS value of and AC waveform is 115 amps. Find the peak value.

Solution:
The peak value of the AC waveform is the RMS value multiplied by the square root of 2.

$$I_{\text{peak}} = I_{\text{RMS}} \times \sqrt{2} = 162.6$$

Look at Figure 3–6. When AC voltage, or current, is rectified, the resulting waveform is not equal to a constant or steady DC. Rectified AC voltage, or current, varies or ripples; however, it never falls below zero. The **average value** of this ripple, or variation, is a steady DC value as measured by a DC meter across the load. The average value of a rectified wave can be calculated by multiplying .637 times the peak value of the rectified wave. For example, if the AC peak rectified value were 110 VAC, then the average value would be 110 × 0.637 = 70.07 V.

3.3 Eddy Currents

When AC flows in a conductor, the alternating current flowing through the conductor causes certain voltages to develop within the conductor.

FIGURE 3–6   Peak and average values.

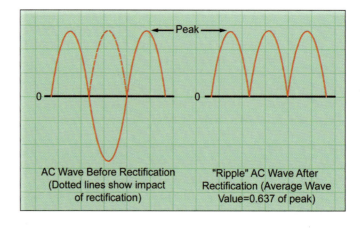

AC Wave Before Rectification
(Dotted lines show impact
of rectification)

"Ripple" AC Wave After
Rectification (Average Wave
Value=0.637 of peak)

These voltages create small internal currents to flow. The name given to this current flow is **eddy current**. While these internally induced currents are not flowing from one end of the conductor to the other, they are nevertheless flowing or moving internally within the conductor. Since any current flow is the result of the flow of free electrons, eddy currents are also using the free electrons to move within the conductor. Because some of the free electrons are used, fewer free electrons are available for use by the current flowing through the conductor from one end to the other. The net effect of eddy current is power loss within the conductor. This power loss increases the overall resistance of the conductor.

3.4 Skin Effect

In a DC circuit, the electrons travel evenly through the entire cross section of the conductor. However, in an AC circuit conductor, besides setting up eddy currents, the voltage that creates the eddy current also causes the current flow in the conductor to be repelled away from the center of the conductor toward the outside of the conductor. The current is forced to travel near the surface of the conductor. This effect, known as the **skin effect,** creates the same consequence as reducing the cross-sectional area of the conductor because the electrons are forced to flow in a smaller area concentrated near the surface of the conductor. The skin effect also causes an increase in the conductor resistance in the circuit due to power losses. Both eddy currents and the skin effect are directly related to the frequency of the circuit. Therefore, as the frequency increases, the magnitude of the eddy currents increases, causing the skin effect to also increase.

Generally, the effects of eddy currents and the skin effect do not have a critical negative impact on circuits except at higher frequencies. The use of stranded cables, which provide more surface area than a solid conductor of an equal size, also helps reduce the overall negative effect of these two elements within the AC circuit.

3.5 Frequency and Wavelength

The number of cycles completed in a second is called **frequency** (f) and is measured in hertz (Hz). Look at Figure 3–7. Notice that the two waveforms have the same peak value (amplitude) but different frequencies. When measured in audio and higher ranges, waveform frequencies are called kilo-, mega-, or gigahertz per second:

- 1,000 cycles per second = 1×10^3 Hz = 1 kHz (kilohertz)
- 1,000,000 cycles per second = 1×10^6 Hz = 1 MHz (megahertz)
- 1,000,000,000 cycles per second = 1×10^6 Hz = 1 GHz (gigahertz)

The amount of time it takes for one cycle is called a **period** (T). The period is equal to the reciprocal of the frequency:

$$T(\text{s}) = \frac{1}{f\,(\text{hertz})} \tag{3.1}$$

FIGURE 3–7 Frequency and amplitude.

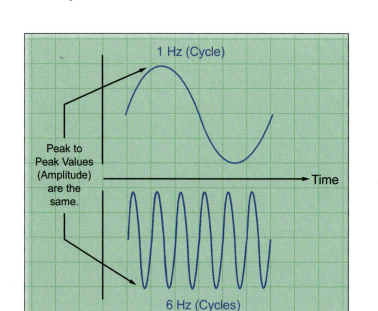

For example, household current is 60 Hz; consequently, the period (T) for one cycle is 1/60 of a second. Because of the short time associated with periods, they are usually measured in the following units:

- T = 1 millisecond = 1 ms = 1×10^{-3} s
- T = 1 microsecond = 1 μs = 1×10^{-6} s
- T = 1 nanosecond = 1 ns = 1×10^{-9} s

The distance that an electromagnetic wave will travel as it completes one cycle is called the *wavelength*. Knowing the frequency and the speed of light (3×10^8 m/s) makes it possible to calculate the wavelength using the formula

$$\lambda = \frac{c}{f} \qquad (3.2)$$

Where:

λ = wavelength in meters or miles
c = speed of light in meters per second or miles per second
f = frequency in hertz.

EXAMPLE 3

Find the wavelength of European household current at 50 Hz.

Solution:
First solve in meters:

$$\lambda = \frac{c}{f} = \frac{3 \times 10^8}{50} = 6 \times 10^6 \text{ meters}$$

Now solve in miles:

$$\lambda = \frac{c}{f} = \frac{186,000}{50} = 3,720 \text{ miles}$$

■ THE MEASUREMENT OF POWER

3.6 Volt-Amperes, Watts, and Volt-Amperes Reactive

You should recall that power is the rate at which energy is being expended. When you studied DC circuits, you learned the power formula, $P = I$E. In AC circuits, the problem of calculating power becomes a little more complex. In fact, there are three types of power in an AC circuit:

1. **True power.** This is the rate at which energy is being used in an electrical circuit and is measured in watts. The energy associated with true power is used by the circuit and not returned to the source. **True power** is also called *real power*.

2. **Reactive power.** This is the rate at which energy is being stored by magnetic and electrostatic fields and is measured in **volt-amperes reactive (VAR)**. The energy associated with reactive power is borrowed by the system inductors and capacitors for one-fourth of each cycle and returned during the next one-fourth of a cycle. **Reactive power** is sometimes called *imaginary power*.

3. **Apparent power.** This is the vector sum of the true power and the reactive power and is measured in volt-amperes. **Apparent power** represents the total rate at which energy is being used and borrowed by the electrical system components.

Figure 3–8 shows the relationship among **watts** (w), VAR, and **volt-amperes** (VA). Recall from your earlier study of right triangles and vectors the following key points:

1. VA is the vector sum of W and VAR.
2. W divided by VA is equal to the cosine of θ.
3. VAR divided by VA is equal to the sine of θ.

3.7 Phase Angle

Phase angle is an extremely important concept in AC circuits. In DC circuits and AC circuits that have only resistance, the voltage and the current are always exactly in phase. That means that they hit their peak values and their zero values at exactly the same time.

FIGURE 3–8   The power triangle.

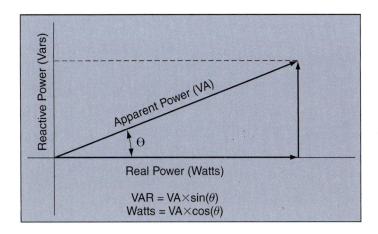

VAR = VA×sin(θ)
Watts = VA×cos(θ)

FIGURE 3–9   Three out-of-phase sine waves.

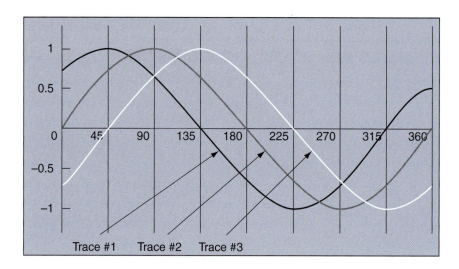

In later chapters, you will learn that the voltage and current are not in phase in an AC circuit with inductors or capacitors (see Figure 3–9).

You know that one full cycle of a sine wave is exactly 360 electrical degrees in length. This concept comes from the fact that a simple AC generator creates one full sine wave for each full rotation of the rotor. Figure 3–9 shows three different waveforms that are out of phase with each other. Notice that trace 1 hits its peaks 45° ahead of trace 2. Also notice that trace 3 hits its peaks 45° after trace 3. In electrical terminology, trace 1 leads trace 2 by 45°. In a similar manner, trace 3 lags trace 2 by 45°. If trace 3 is the current in a circuit and trace 2 the voltage, then the current lags the voltage by 45°.

3.8 Power in an AC Circuit

In Figure 3–8, the angle shown as θ is significant for two reasons:

1. θ is the angle between the true power (watts) and the apparent power (VA).
2. θ is also the phase angle by which the current lags (or leads) the voltage in the circuit.

This leads to two very important concepts:

1. If θ = 0, VAR = 0, and watts are maximum. This is true because the cos(0) = 1.
2. If θ = 90, VAR = maximum, and watts = 0. This is true because the sin(0) = 0.

A lightbulb is almost pure resistance, which means that the applied voltage and the resulting current are in phase (see Figure 3–10). From the previous discussion, you can see that this means that a lightbulb converts nearly all the applied volt-amperes to watts.

FIGURE 3–10   Current and voltage in phase create pure watts.

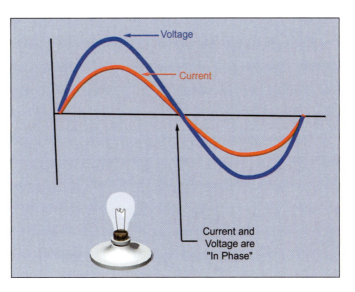

■ SUMMARY

In this chapter, you learned that the sine wave is used to show the output for a generator rotating at a constant speed. You also learned that DC was the original source of electricity but that AC is more useful, as it can be transformed easily to different voltage levels. A transformer permits voltage to be stepped up or down. The advantage of high-voltage transmission is that more power can be distributed at the same current or the same power can be distributed through smaller wire.

Alternating current reverses its direction of flow at periodic intervals. There are 360° in one complete sine wave. Each sine wave has two alternations—one positive and one negative. One complete sine wave or waveform is called a cycle. The number of complete cycles that occur in a second is called the frequency. Frequency is measured in hertz (Hz).

The instantaneous value can be defined as a value found at any point or instant along the waveform. The peak-to-peak voltage is the amount of AC voltage measured from the positive peak to the negative peak. The peak value is the maximum instantaneous value of voltage attained in the waveform. The RMS voltage, also referred to as the effective voltage, will produce as much power as a like amount of DC voltage.

The average voltage can be calculated by averaging all the voltage values during any cycle of the sine wave. If the waveform is a perfect sine wave, the average value is zero. A full-wave rectified sine wave (see Figure 3–6) has the same RMS value as a sine wave; however, the full-wave rectified sine wave has a DC (average) value equal to 63.7% of the peak.

■ REVIEW QUESTIONS

1. Discuss the concepts of RMS, average, peak, and peak to peak.
   a. How are they calculated for a sine wave?
   b. How are they calculated for a pure DC?
   c. How are they calculated for a sine wave that is not centered at zero?
2. How are period and frequency related to each other?
3. How are wavelength and frequency related to each other?
4. A two-pole generator (like the one shown in Figure 3–3) is rotating at 3,600 revolutions per minute (rpm). What is the generator output frequency in hertz? Can you use this information to develop a formula for output frequency of such a generator versus the rpm?

5. Discuss the skin effect.

   a. What are its implications for the efficiency of an AC circuit?

   b. A certain wire size will carry 100 amperes of DC without overheating. By comparison, how much AC current would the same wire carry?

6. Discuss the relationship among apparent power, true power, and reactive power.

■ PRACTICE PROBLEMS

1. Frequency is measured in ____, which is cycles per second.

2. The time it takes to complete one cycle is called a ____.

3. If the period is 1 microsecond, what is the frequency?

4. Twelve cycles of a sine wave occur in 3 milliseconds. What is the frequency?

5. The sine of an angle is maximum positive when the angle is ____ degrees.

6. If a sine wave has a peak value of 5 volts, what is its
   a. average value?
   b. effective (RMS) value?

7. What is the peak-to-peak value of a sine wave having an average value of 5 volts?

8. In a 120-V circuit, the current of 5 amperes is lagging the voltage by 30°. Calculate the following values:
   a. Apparent power
   b. True power
   c. Reactive power

9. The apparent power in a certain circuit is 1,000 VA. Answer the following questions (do not use a calculator).
   a. If the voltage and current are exactly in phase, what are the true power and the reactive power values?
   b. If the voltage and current are exactly 90° out of phase, what are the true power and the reactive power values?

chapter 4

AC Resistive Circuits

OVERVIEW

With the exception of the skin effect and eddy currents (discussed in Chapter 3), the analysis of AC resistive circuits is identical to the analysis of DC resistive circuits. This chapter looks at resistive AC circuits. Fortunately, the skin effect and eddy currents are very small at power frequencies up to 60 hertz. In DC circuits, there is only one type of load: resistive.

This means that all the techniques that you learned in your DC course will apply to AC resistive circuits. A few items do need to be reviewed. Most of these you have studied already, but they are reviewed here.

OBJECTIVES

After completing this chapter, you should be able to:

1. Analyze an AC resistive circuit.
2. Draw the current and voltage sine wave for an AC resistive circuit.
3. Determine instantaneous values in an AC resistive circuit.

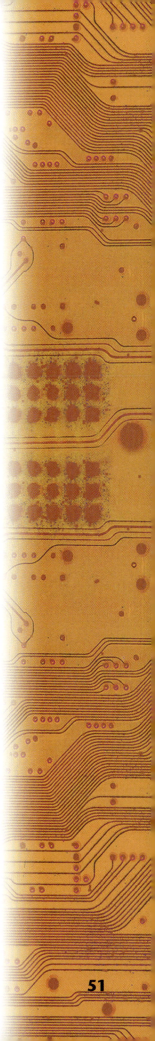

■ RESISTIVE LOADS

Resistive loads are loads that contain pure resistance, such as lighting and heating elements. Resistive loads are characterized by two factors: they produce heat, and the current and voltage are in phase with each other.

In a DC circuit, only resistance opposes current flow. This is not true in AC circuits, which have other factors, such as capacitance and inductance. Such factors can usually be ignored in a resistive circuit when making calculations because they are so small.

Whenever current flows through a resistance, heat is produced. That is why a wire becomes warm when current flows through it. The heating elements of an electric stove and a lightbulb become extremely hot because of resistance.

4.1 Voltage and Current Phasing

Look at Figure 4–1. When an AC voltage is applied to an element with only resistance (such as the lightbulb in the figure), the sine wave representation of the current flow will mimic (follow "in phase," no shift or delay time) the sine wave representation of the voltage and will reverse direction when the voltage reverses polarity. We say that "the current is in phase with the voltage." Note that the current will rise and fall proportionally with the voltage.

4.2 Instantaneous Value of a Sine Wave

As you learned earlier, the AC waveforms used in electrical power applications are in the form of sine waves. This means that any instantaneous value on the wave can be determined by using the formula

$$I_{inst} = I_{peak} \times \sin(\theta) \text{ and } E_{inst} = E_{peak} \times \sin(\theta) \tag{4.1}$$

FIGURE 4–1   Voltage and current in phase.

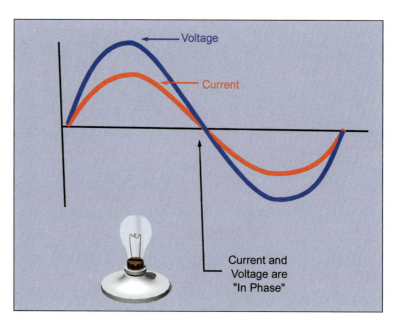

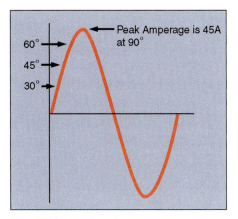

FIGURE 4–2   Instantaneous sine wave values.

Look at Figure 4–2. You can find the instantaneous value by multiplying the peak voltage, or current, times the sine of the angle at this specific point (a specific point on the sine wave at a specific point in time) at which you want the measurement. The formula for this calculation is given next.

EXAMPLE 1

What is the instantaneous current value of a 45-ampere (peak) AC circuit at 30°, 45°, and 60°?

Solution:

$$I_{inst} = 45 \times \sin(30) = 45 \times 0.5 = 22.5 \text{ V}$$
$$I_{inst} = 45 \times \sin(45) = 45 \times 0.707 = 31.8 \text{ V}$$
$$I_{inst} = 45 \times \sin(60) = 45 \times 0.866 = 38.97 \text{ V}$$

■ VECTORS

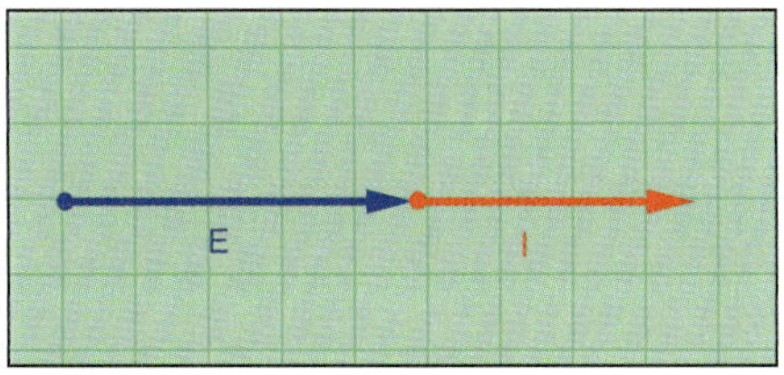

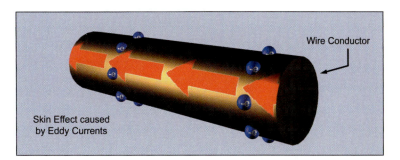

FIGURE 4–3   Vectors.

A vector is a line that indicates both direction and magnitude and is represented by an arrow. Its length indicates the magnitude, and the direction is indicated by a counterclockwise angle measured from 0°.

Look at Figure 4–3. Zero degrees is indicated by a horizontal line. We can show voltage and current together on a graph. Does this mean that we can add these two together? No. You cannot mix apples and oranges, but it does show they are in phase (going the same direction at the same time).

■ RESISTIVE FACTORS IN AN AC CIRCUIT

4.3 Skin Effect

Refer to Figure 4–4. In a DC circuit, as the electrons travel through a wire, they are distributed evenly across the entire cross section. If that same conductor were connected to a source of AC, the resistance to the flow of current would be slightly higher because of the skin effect. The alternating current sets up eddy currents in the conductor. These currents cause the electrons to be magnetically repelled toward the outer surface of the conductor. As a result, there is less cross-sectional area in which the AC current can flow as compared to an equivalent amount

FIGURE 4–4   Skin effect.

of DC current when using the same-size conductor. The wire acts as if it is a piece of tubing with no conductor material in the middle to provide for current flow. Since there is less conductive material available for current flow, the resistance to current flow increases.

The skin effect is proportional to the frequency because it is the frequency that causes the effect. (The skin effect is a product of self-inductance, which is affected by frequency.) Therefore, as frequency increases, skin effect increases.

◼ AC SINE WAVE VALUES

There are four forms of measuring current and voltage associated with AC sine waves: peak, peak to peak, RMS, and average.

4.4 Peak and Peak-to-Peak

The peak value is measured from zero to the highest value obtained in either the positive or the negative direction. Therefore, the peak-to-peak value is twice that of the peak because it is measured from one peak value to the other.

4.5 RMS

The RMS value is also called the *effective value*. Consider Figure 4–5. The blue trace (trace 1) is a DC voltage equal to 120 V. If this voltage

FIGURE 4–5   The sine wave and its RMS value.

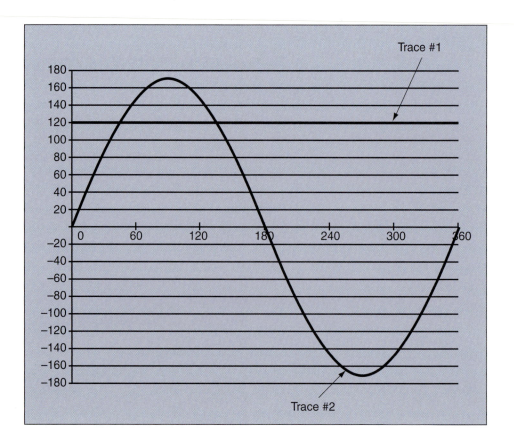

were applied to a 100-Ω resistor, the power dissipated could be calculated using the familiar power formula

$$P = \frac{E^2}{R} = \frac{(120)^2}{100} = 144 \text{ W}$$

The question arises, What is the peak value of an AC sine wave that will dissipate the same amount of power 120 VDC? Imagine connecting an AC voltage source and a wattmeter to the 100-Ω resistor. Now gradually increase the AC voltage until the same 144 W are being dissipated. For a sine wave, this will occur when the peak voltage is at $120 \times \sqrt{2} = 169.71$ V. Put another way, 120 V is the effective, value of an AC voltage with a peak of 169.71 V.

This value can be calculated mathematically by taking the square root of the average of the square of the AC waveform. This is where the term *RMS* originates. For a sine wave, the RMS value is equal to the peak value multiplied times the square root of 2.

EXAMPLE 2

A circuit has a peak value of 423 amps. Find the RMS value.

Solution:

$$I_{RMS} = \frac{I_{peak}}{\sqrt{2}} = \frac{423}{\sqrt{2}} = 299 \text{ A}$$

EXAMPLE 3

A circuit has an RMS value of 115 amps. Find the peak value.

Solution:

$$I_{peak} = I_{RMS} \times \sqrt{2} = 115 \times \sqrt{2} = 162.6 \text{ A}$$

4.6 Average

The average value does not apply to a pure AC sine wave. It is used primarily when an AC sine wave is changed to a rippling DC waveform by a full-wave rectifier. Recall that the effective value causes the same amount of heat throughout both the positive and the negative half cycles as the same value of DC. The average value is the actual average of all the instantaneous voltages or currents values across a full cycle. This is the reason that a pure sine wave has no DC value: it is above the zero exactly as much as it is below. Therefore, a pure AC sine wave's average is zero.

A rectified sine wave, on the other hand, has most or all of its wave form above zero. The actual DC value depends on whether it is a full-wave rectifier, as in Figure 4–6, or a half-wave rectifier, as in Figure 4–7.

FIGURE 4–6   Average value of a full-wave rectifier.

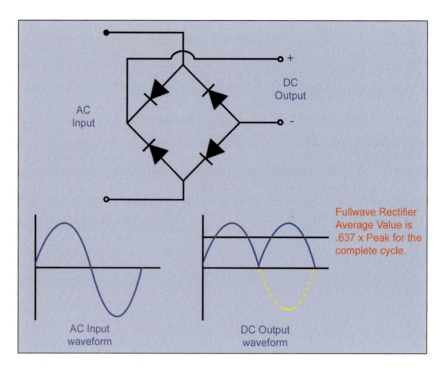

FIGURE 4–7   Average value of a half-wave rectifier.

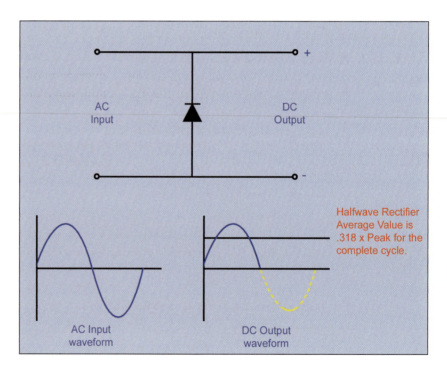

The formula for calculating the average value of a full-wave rectified wave is calculated mathematically as

$$V_{\text{AVE}} = \frac{2 \times V_{\text{peak}}}{\pi} = \frac{2 \times V_{\text{RMS}} \times \sqrt{2}}{\pi} \tag{4.2}$$

Note that a shortcut can be used for the previous formula based on the fact that $(2 \times \sqrt{2})/\pi \approx 0.9$. Thus, multiplying the RMS value of the full-wave rectified sine wave by 0.9 gives the average value.

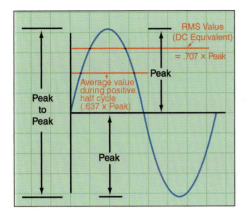

FIGURE 4–8   Peak, peak-to-peak, RMS, and average values.

EXAMPLE 4

The effective value of a full-wave rectified sine wave is 30 A. Find the average value.

Solution:

$$V_{AVE} = \frac{2 \times V_{RMS} \times \sqrt{2}}{\pi} = \frac{60 \times \sqrt{2}}{\pi} = 27\ A$$

You might guess that the average value of a half-wave rectified sine wave would be equal to half the value of the full-wave rectified sine wave. You would be right. The formula for calculating the average value of a half-wave rectified sine wave is

$$V_{AVE} = \frac{V_{peak}}{\pi} = \frac{V_{RMS} \times \sqrt{2}}{\pi} \tag{4.3}$$

■ SUMMARY

In this chapter, you learned that the instantaneous voltage at any point on a sine wave is equal to the peak, or maximum voltage, times the sine of the angle of rotation at that instant. The peak-to-peak voltage is the amount of voltage measured from the positive-most peak to the negative-most peak. The peak value is the maximum amount of voltage of the sine wave form, and since the positive peak is equal to the negative peak, $V_{peak\ to\ peak} = 2 \times V_{peak}$.

The RMS value of voltage will produce as much power dissipation as a like amount of DC voltage, and the average value (current or voltage) applies when AC is rectified to DC by either a half-wave or a full-wave rectifier. Refer to Figure 4–8. Note that the average value shown in the figure applies *only* if the sine wave has been rectified by a full-wave rectifier.

The current and voltage of a purely resistive circuit are in phase with each other. Resistance in AC circuits is characterized by the fact that the resistive part will produce heat, such as in a lightbulb.

Electrons are forced toward the outside surface of a conductor because of eddy currents produced as a result of the frequency of the alternating current. This action, called the skin effect, increases the conductor's resistance and is directly proportional to the frequency.

■ REVIEW QUESTIONS

1. A 120-W resistance heater is being used in a cabinet. It is rated at 120 VAC. If you wished to use DC for the heater, how much voltage would you apply? Why?

2. A certain wire size is rated for 100 amperes AC at 60 hertz. Will the wire be capable of carrying more or less current at the following frequencies?

   a. DC

   b. 400 Hz

   c. 20,000 Hz

3. A lightbulb is connected to a 120-VAC supply. If an inductor is placed in series with the lightbulb, will it glow brighter or less bright?

4. Why is the effective value of a half-wave rectified sine wave less than the effective value of a full-wave rectified sine wave? Why so for the average value?

5. Why is the average value of a sine wave equal to zero?

Three-Phase Systems

■ **OUTLINE**

PHASES AND PHASE ANGLES
THREE-PHASE PRODUCTION

WYE-CONNECTED SYSTEMS
DELTA CONNECTIONS

■ OVERVIEW

This chapter presents the technical concepts of electrical power that use multiple, equal voltages separated by some phase angle. Such systems are called **polyphase** systems.

Although research has been performed with polyphase systems of up to 12 individual voltages, most of the electrical power generated and used by industry in the world today is three phase. This situation has existed since the early twentieth century, when the technical and economic advantages of three-phase systems became apparent.

In your career, you will encounter a wide variety of single-phase and polyphase systems. The material in this chapter provides the basic information that you will need to understand, work with, and analyze a wide variety of different schemes.

■ OBJECTIVES

After completing this chapter, you should be able to:

1. Explain how single-phase, two-phase, and three-phase power is generated.
2. Describe the fundamental characteristics of single-phase and polyphase systems.
3. List the advantages of a three-phase system over a single-phase system.
4. Calculate currents and voltages in a three-phase power system.

■ GLOSSARY

φ The Greek letter phi (pronounced "*fi*"). The symbol used as a shorthand for the word *phase.*

Delta connection Connecting the windings of a generator or transformer so that each end of each winding is connected to one end of each of the other winding. The three junction points become the phase connections. The name is derived from the fact that, when properly oriented, the winding connections resemble the uppercase Greek letter delta (Δ).

Polyphase A method of transmitting electrical power by using multiple wires. The generated voltages are equal in magnitude but separated in phase angle.

Single-phase An electrical distribution scheme that employs only one source voltage. Abbreviated as 1 φ.

Three-phase A polyphase system made up of three equal voltages separated by a phase angle of 120°. Abbreviated as 3 φ.

Two-phase A polyphase system made up of two equal voltages usually separated by a phase angle of 180°. A few two-phase systems use a 90° separation. Abbreviated as 2 φ.

Wye connection Connecting the windings of a generator or transformer so that the three windings are connected together at one end. The opposite ends of each winding become the phase connections. The name is derived from the fact that, when properly oriented, the winding connections resemble the letter "Y."

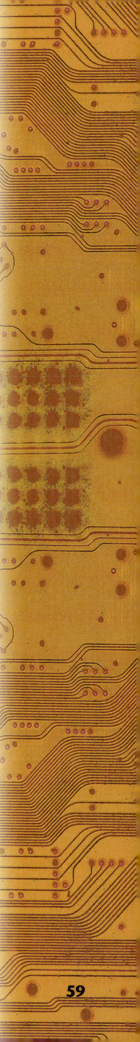

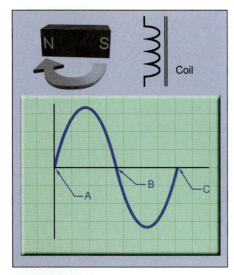

FIGURE 5–1   Single-phase voltage generation.

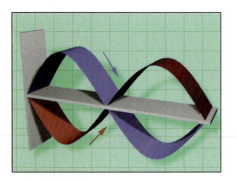

FIGURE 5–2   Two voltages that are 180° out of phase.

■ PHASES AND PHASE ANGLES

Rotating a magnetic field through the conductors of a stationary coil generates a **single-phase** alternating voltage (see Figure 5–1). As the alternating poles of the magnet cut through the conductors of the stationary coil, the voltage induced will alternate at the same speed. The output of such a generator is called *single phase* because it has only one voltage.

Although simple in design and concept, single-phase generation and distribution has some draw backs:

1. During the complete rotation of the magnet through the conductors of the stationary coil, the voltage produced drops to zero three times (points A, B, and C in Figure 5–1). This causes the resulting power from the supply to pulse from maximum to zero during each cycle.

2. The pulsing of the power output causes vibration and excessive physical stress on the generator. Twice each cycle, the generator sees the maximum magnetic field, and twice each cycle, it sees the minimum.

3. Since the generator output varies with each cycle, its efficiency is low.

As you learned in Chapter 3, when two voltages are in phase with each other, they have the same frequency, and there is no time displacement between them. Both voltages start at the same point on the baseline and cross the baseline at the same time, in the same direction. If two voltages cross the baseline at the same time going in opposite directions yet have the same frequency, they are said to be 180° out of phase. Figure 5–2 is a three-dimensional graph of two voltages that are 180° out of phase.

Although **two-phase** systems have been used (and still are) to some extent, the **three-phase** system has become the power configuration of choice. Figure 5–3 shows a three-phase, 480-volt system similar to the many that you will encounter in your career as an electrician. Notice that the three voltages labeled A, B, and C are the same magnitude (277 V_{RMS}) and are 120° out of phase. The A phase is almost always used as the reference, with the B phase lagging by 120° and the C phase by 240°.

Figure 5–4 shows why such a system is called a 480-V system. In the figure, 5 Aφ is drawn as a vector (V_{AN}). Since it is the reference, it is drawn horizontally to the right in the 0° position. The length of the vector is equal to the RMS value of 277 V. The vector could be drawn using the peak voltage of 391 V; however, universally the industry uses the RMS values for such drawings.

The B phase is drawn in the 120° lagging position and the C phase in the 240° lagging position. Using the vector diagram makes the relationship among the three-phase voltages clearer. The A, B, and C phase voltages are actually the voltages measured from each wire to ground or neutral. If the voltage from, for example, the A to the B phase is measured, it is equal to the length of one side of an equilateral triangle, as shown in the illustration. Geometrically, the length of the side of the

FIGURE 5–3   A three-phase
voltage system.

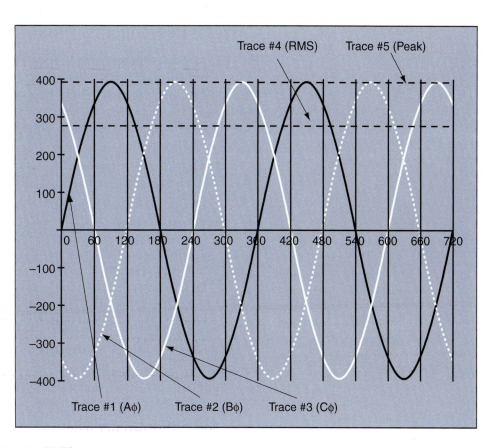

Trace #4 (RMS)   Trace #5 (Peak)

Trace #1 (Aφ)   Trace #2 (Bφ)   Trace #3 (Cφ)

FIGURE 5–4   A vector drawing
of a three-phase system.

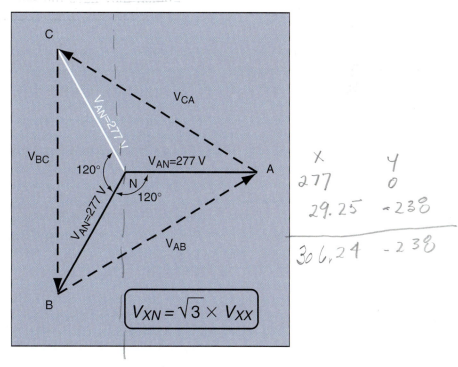

$$V_{XN} = \sqrt{3} \times V_{XX}$$

equilateral triangle is equal to the square root of 3 multiplied by one of
the neutral voltages:

$$V_{\mathrm{AB}} = \sqrt{3} \times V_{\mathrm{AN}} = \sqrt{3} \times 277 = 480 \text{ V} \qquad (5.1)$$

This $\sqrt{3}$ relationship is true for all three-phase systems.

■ THREE-PHASE PRODUCTION

Three-phase voltages are generated by placing three coils separated by 120° on the stator of the generator. As the rotor turns, it cuts through each of the coils 120° apart. This is how three-phase voltage is produced. Figure 5–5 shows how this is done.

Since the rotor's magnetic lines cut through each of the coils 120° apart, the three sine waves produced will be 120° out of phase with each other. If the output of the generator is connected to a three-phase load (each wire is connected to the same resistance), the output power of the generator will be constant during its entire 360° rotation. As the output from one phase drops off, the output of the other two will begin to pick up; consequently, the output power remains constant. Figure 5–6 is similar to Figure 5–3 except the phases are now labeled as A phase, B phase, and C phase.

Three-phase circuits have many advantages:

1. The horsepower rating of three-phase motors and the kilovolt-amp rating of three-phase transformers are 150% greater than for single-phase motors and transformers of the same physical size.

2. At no time does the voltage to the load ever drop to a zero level. Since the voltage to the load never drops to zero, the power consumed by the load never goes to zero. In fact, the total power from the generator is constant, as discussed previously.

3. The conductors for three-phase systems need be only 75% the size of conductors for a single-phase two-wire system.

FIGURE 5–5 Coils 120° apart for three-phase generation.

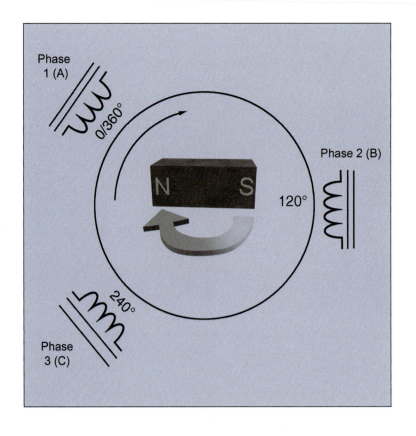

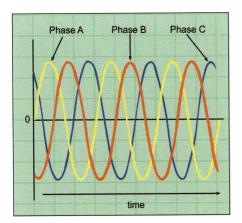

FIGURE 5–6   Three-phase voltage generation.

4. There are two methods of connecting the wires in a three-phase system. These two approaches offer further variation in the system's use and application. How the generators are connected will determine the system characteristics.

■ WYE-CONNECTED SYSTEMS

The **wye connection** is made by connecting one end of each of the three-phase windings together (see Figure 5–7). The wye system has the advantage of providing two sets of three voltages: the three phase-to-neutral voltages and the three phase-to-phase voltages.

Figure 5–7 also shows the way that voltages are measured in a wye-connected system. The voltage measured across a single winding of the phase is known as the phase voltage, phase-to-ground voltage, or phase-to-neutral voltage. The voltage measured between the lines is known as the line voltage, the line-to-line voltage, or the phase-to-phase voltage.

To calculate the line voltage in a wye-connected system, remember that the line voltage is higher than the phase voltage by a factor of the square root of the number of phases (three). This is because the voltage of each phase has a 120° delay or shift that has to be taken into account when the voltages are measured.

In a three-phase wye-connected system, the voltages are related by

$$E_{phase} = \frac{E_{line}}{\sqrt{3}} \qquad (5.2)$$

In a three-phase wye-connected system, phase current and line current are the same because they are in series:

$$I_{line} = I_{phase} \qquad (5.3)$$

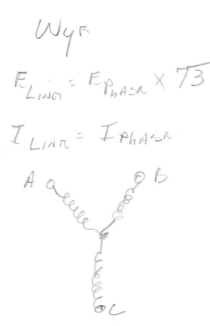

FIGURE 5–7   Wye connection and voltage meaurements.

Line Voltage 208V

Phase Voltage 120V

EXAMPLE 1

Refer to Figure 5–8. The phase-to-neutral voltage of a system is 240 V. Each phase has a 30-amp load. Find the current (I) of each line of the wye connection. Find the voltage (E) across lines B and C. What is the power delivered to each load?

Solution:

$$I_{\text{line}} = I_{\text{phase}} = 30 \text{ A}$$

$$E_{\phi\phi} = \sqrt{3} \times E_{\phi N} = 1.732 \times 240 = 415 \text{ V}$$

Assuming the voltage and current are in phase with each other, the power delivered to each load is

$$P = E_{\phi N}I_{\phi} = 240 \times 30 = 7{,}200 \text{ W}$$

The power delivered to all three loads will be three times that amount:

$$P_{3\phi} = 3 \times E_{\phi N} \times I_{\phi}$$

Noting that

$$E_{\phi N} = \frac{E_{\phi\phi}}{\sqrt{3}}$$

substituting that into the equation above it yields

$$P_{3\phi} = 3 \times \frac{E_{\phi\phi}}{\sqrt{3}} \times I_{\phi} = \sqrt{3} \times E_{\phi\phi} \times I_{\phi} = 21.56 \text{ kW}$$

Note the rounding error when using the approximation for the square root of 3.

FIGURE 5–8   Wye current measurements.

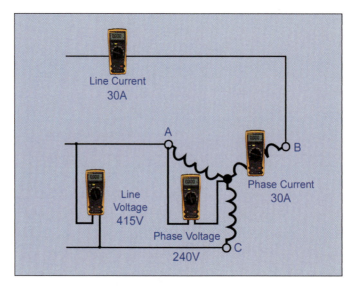

■ DELTA CONNECTIONS

Three-phase power systems can also be connected in a delta configuration. This connection receives its name because the schematic diagram resembles the Greek letter delta (Δ). The **delta connection** is made by connecting the end of winding 1 to the beginning of winding 2, connecting the end of winding 2 to the beginning of winding 3, and connecting the end of winding 3 to the beginning of winding 1. More simply stated, the delta connection is made by connecting the three windings end to end. Figure 5–9 is an example of a delta-connected system.

In a delta connection, line voltage and winding voltage are the same. In delta-wired systems, line voltage is equal to winding voltage. The line current and winding current, however, are different. The line current of a delta connection is higher than the winding current by a factor of $\sqrt{3}$ (1.732). This can be proven mathematically by calling on Kirchhoff's current law and your vector addition skills. Consider Figure 5–10.

EXAMPLE 2

For Figure 5–10, calculate the line currents (I_A, I_B, and I_C) if the winding currents are as shown. Use a scientific calculator to perform the vector calculations.

Solution:
This is readily solved by applying Kirchhoff's current law at each node. For the top node (Aϕ),

$$I_A = I_{AB} - I_{CA} = 1\angle 30 - 1\angle 150 = 1.732\angle 0$$

For the bottom right node (Bϕ),

$$I_B = I_{BC} - I_{AB} = 1\angle -90 - 1\angle 30 = 1.732\angle -120$$

For the bottom left node (Cϕ),

$$I_C = I_{CA} - I_{BC} = 1\angle 150 - 1\angle -90 = 1.732\angle 120$$

[handwritten annotations:]
.8660   .5
− .3660   .5
1.732   1

FIGURE 5–9 Delta three-phase connection.

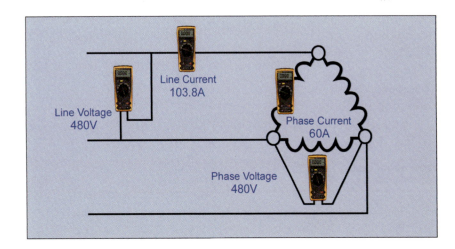

Line Current 103.8A

Line Voltage 480V

Phase Current 60A

Phase Voltage 480V

FIGURE 5–10 Calculating line currents in a delta system.

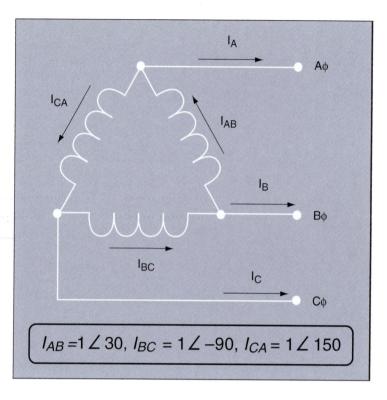

$$I_{AB} = 1\angle 30, \; I_{BC} = 1\angle -90, \; I_{CA} = 1\angle 150$$

A couple of points should be made:

1. The winding current value angles were chosen so that the line currents would have the familiar $0°$, $-120°$, and $120°$ angles. Note that the winding currents are, in fact, $120°$ apart.

2. Using a magnitude of 1 for the winding currents allows the answer to be very clearly equal to $\sqrt{3}$ (1.732). This means that whatever the actual magnitude is in the windings, the line currents will have a magnitude that is 1.732 times greater.

■ SUMMARY

In the early days, electrical power was generated and distributed as DC or single-phase AC. For a variety of reasons, these systems have been gradually supplanted with three-phase systems.

The voltages of a three-phase system are $120°$ out of phase with each other. The two types of connections used with three-phase systems are wye and delta connections. Wye connections have one end of each of the windings connected together. In a wye connection, the winding current and the line current are the same (acting as series current). The phase voltage is less than the winding voltage by a factor of 1.73.

Delta connections have the windings connected end to end. In a delta connection, the phase voltage and the line voltage are the same. The phase current is less than the line current by a factor of 1.73 and acts as parallel current (more than one current path).

■ REVIEW QUESTIONS

1. Discuss the way in which a generator must be built to deliver the following:
   a. Single-phase power
   b. Two-phase power
   c. Three-phase power
   d. Polyphase power with any number of phases

2. List and discuss at least three advantages of three-phase power over single-phase power.

3. What are the two types of connections used for three-phase power systems? What are the advantages and disadvantages of each?

4. Discuss and explain the following:
   a. Magnitude relationships in a three-phase system (phase to phase versus phase to ground)
   b. Angle relationships in a three-phase system

■ PRACTICE PROBLEMS

1. Two waves having the same frequency but passing through zero at different times are said to be out of ____.

2. A wye-connected generator delivers 4,160 volts line to line. What is the phase (coil) voltage?

3. If the total current in question 3 is 40 amps, what is the total power?

4. A delta transformer has 30 amps per phase. What is the line current?

PART
2

AC AND DC GENERATORS

chapter 6

DC Generators

■ OVERVIEW

Although AC is available at most sites, DC current has applications that AC cannot support. DC can be used for precise speed control of motors and for large DC motors found in many industrial and manufacturing applications. DC motors and generators are also used in locomotives. Where DC is required, electricians will find DC generators.

As variable frequency drives become more common, the use of DC will become less required. Eventually, DC will be found only in those places where extremely high torque per horsepower is required (motors) or where the nature of the work requires DC, such as in electrolysis applications.

This chapter teaches you how to identify the major parts of a DC generator and explains the fundamental operating principles. Remember that DC machines will be with us for many more years, and knowledge of how they operate and how to install and repair them will be a valuable skill.

■ OBJECTIVES

After completing this chapter, you should be able to:

1. Identify the major parts of a DC generator.
2. Describe the principles of operation of the DC generator.

■ GLOSSARY

Armature   The part of the generator that is connected to the load. The armature of all DC generators is the rotating part. All armatures have an alternating current in them. The rotating loops of wire that rotate through the field to create an alternating voltage.

Armature reaction   The bending of the field magnetic flux by the armature. Armature reaction causes a displacement of the neutral point with attendant arcing and voltage drop.

Brushes   Sliding contacts, usually made of a carbon or graphite alloy, that are positioned so they are always connected to the same polarity armature segments. The brushes ensure that the output from the AC voltage created on the armature is commutated (rectified) to DC.

Commutator   A multisegment rotating switch that is connected to the armature windings.

Compensating windings   See Interpole windings.

Field   The part of the DC generator that creates the magnetic field that is cut by the armature windings. The permanent magnetic field through which the armature windings move to create the output voltage. On a DC generator, the field is located on the rotor.

Interpole windings   Field windings that are located physically between the main field poles. Interpoles are used to reduce armature reaction and are supplied with armature current.

Thomson-Ryan windings   See Interpole windings.

■ GENERATOR CHARACTERISTICS

6.1 Inducing Voltage

A generator is a device that converts mechanical energy to electrical energy. A voltage is induced in a conductor when it cuts magnetic lines of flux. This principle, called *electromagnetic induction,* is the underlying principle of operation for both DC and AC generators. Figures 6–1 through 6–4 show how this principle works. As the loop is rotated and cuts through the lines of flux, a voltage is induced in the loop. An easy way to remember the relationship between motion, electromagnetic lines of force, and current flow is to use Fleming's left-hand rule for generators, shown in Figure 6–5.

6.2 Sine Wave Generation and Conversion to DC

Look at Figure 6–6. First, look at the **armature** loop at position A at the top of the figure. The position of the wire loop is such that it is moving parallel to the magnetic lines of flux. Now look at the waveform at po-

FIGURE 6–1   Positions A and E.

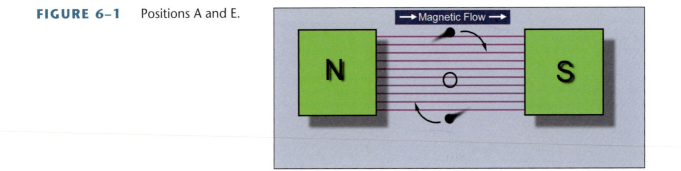

FIGURE 6–2   Position B.

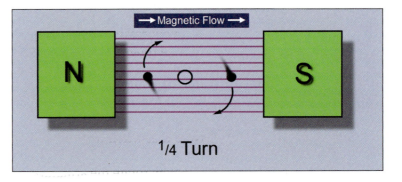

FIGURE 6–3   Position C.

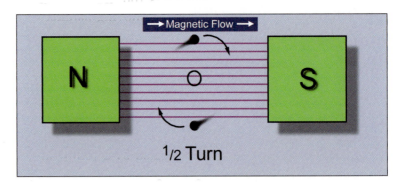

FIGURE 6–4   Position D.

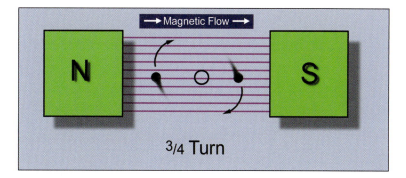

FIGURE 6–5   Fleming's left-hand rule for generators.

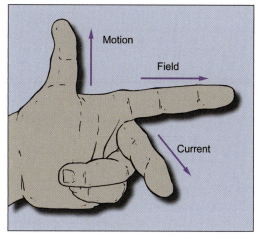

sition A. When the wire loop motion is parallel to the magnetic lines of flux, there is no voltage being produced because the wire loop that makes up the armature is not cutting through any of the lines of flux when the loop is parallel to the lines of flux.

Now rotate the armature from position A to position B. As the armature rotates, the motion of the wire loop goes from being parallel to being perpendicular to the lines of flux. As the armature rotates through this 90° travel, it begins to cut through more and more lines of flux. When the armature gets to the 90° point, it is cutting the maximum number of lines of flux per second. Now look at the waveform at position B. Note that when the armature is cutting through the most lines of flux, the highest amount of voltage is produced in the positive direction.

Now rotate the armature through another 90°. At this point, the armature is 180° from its original starting point. Note again that the armature is moving parallel to the magnetic lines of flux and that the voltage produced is zero.

As the armature continues to rotate through the next 90°, the armature is again approaching the point where it will be moving perpendicular to the lines of flux. Take a look at the waveform and notice that the value has gone negative. This change in polarity is due to the motion of the leading edge of the wire loop relative to the direction of the magnetic lines of flux. Notice that the leading edge of the wire loop is indicated by a yellow highlight. When the leading edge is moving up, the polarity of the voltage being produced is negative.

FIGURE 6–6 One cycle of induced voltage.

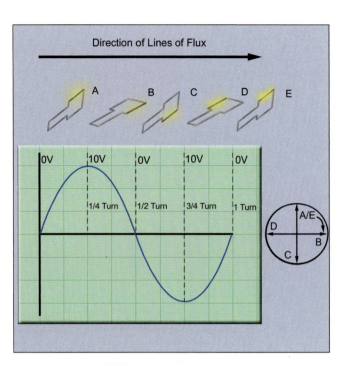

As the loop continues to rotate back to its starting point, the voltage again begins to approach zero. The voltage being produced as this wire loop (the armature of the generator) rotates through a complete 360° of travel through the magnetic lines of flux is called *alternating voltage* because the voltage travels positive for a part of the cycle and then alternates and becomes negative for part of the cycle.

Since the voltage produced by all rotating armatures is alternating voltage, as can be seen in Figure 6–6, the method of removing the voltage from the armature is the key to determining whether the output voltage of the generator is alternating or direct.

6.3 Commutating the Output of an AC Armature

Since DC generators must produce DC current, the method of removing the voltage from the armature is different than that used for an AC generator. In the DC generator, the components used to remove the voltage from the armature are **brushes** and a **commutator** (see Figure 6–7a).

A commutator is a circular ring that is divided into segments that are electrically isolated. Since the commutator is physically connected to the armature loop, the segments have an AC voltage on them. However, in the example generator in Figure 6–7a, the commutator is positioned so that the segment that is on top is always the negative one. This means that the brush on the top will always be connected to the negative brush, and the output voltage from the brushes will be a pulsating DC (see Figure 6–7b).

Unlike Figure 6–6, which shows the voltage produced as being AC, Figure 6–7b shows the resulting waveform of the voltage when the voltage is measured at the brushes. Notice how the output is always positive.

As the armature loop moves from position A to position B, a positive voltage is produced that starts at zero and gradually increases to

FIGURE 6–7 (a) DC generation with a commutator; (b) brush voltage on a DC generator.

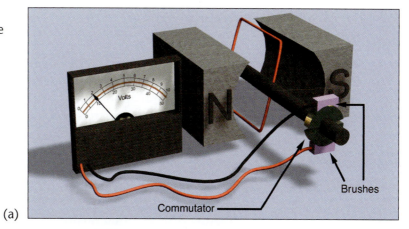

(a)

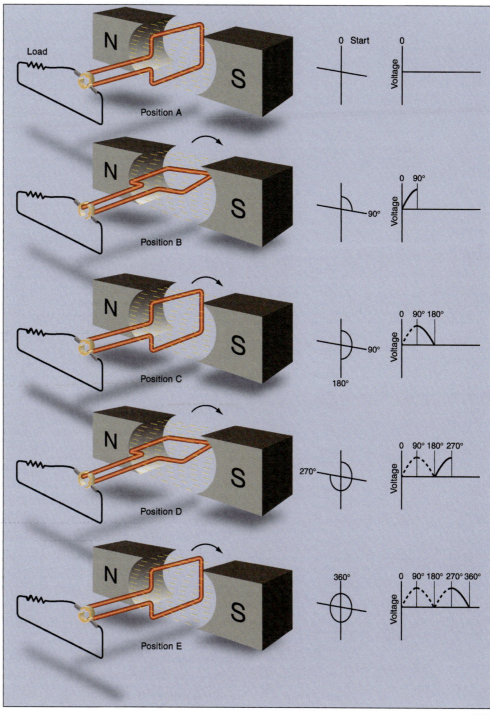

(b)

maximum. As the rotation continues from position B to position C, the loop continues to cut through the lines of flux; however, as the armature position approaches position C, it is producing less positive voltage. When the armature reaches position C, it is moving parallel to the lines of flux, and therefore no voltage is produced.

From position A through position C, the loop produces a positive voltage that has been removed from the armature windings by the brushes that have been riding on the same commutator segments throughout this 180° of armature travel. However, as the armature begins its travel from position C to position D, the brushes will begin to ride on the opposite commutator segments. The change in commutator segments will allow the voltage to remain positive even though the armature is producing an alternating voltage.

As the commutator rotates from position C to position D, the positive lead of the generator is now connected to the segment of the commutator that is producing a positive voltage. At position D, the armature loop will again produce its maximum voltage and then return to zero as the armature rotates to position E.

Regardless of which half of the loop is producing a positive voltage and which half is producing a negative voltage, the commutator and brushes will always cause the output voltage to remain at the same polarity. This is the basic principle of DC generator operation.

■ GENERATOR MAJOR COMPONENTS

6.4 Armature Windings

The rotating member of the DC generator is the armature. The entire assembly includes the iron core, the commutator, and the windings. There are three basic types of windings, depending on the needs of the machine: lap, wave, and frogleg (see Figures 6–8 and 6–9).

1. Lap-wound armatures are used in machines designed for low voltage and high current. Typically, lap windings are designed with large wire to safely handle large currents. When equipment has lap-wound construction, the windings are connected in parallel. This parallel path construction provides multiple paths for current flow. A common characteristic of lap-wound equipment is that there are as many pairs of brushes as there are pairs of poles. Another characteristic of lap-wound equipment is that there are as many armature current paths as there are pole pieces.

2. Wave-wound armatures are used in machines designed for high voltage and low current. Since the current capacity of wave-wound equipment is relatively low, the winding wire size is smaller than that of the lap wound and is typically connected in series so that the voltage of each of the windings is additive. Keep in mind that in series circuits, the current is the same throughout the circuit. The same is true with wave-wound windings. Even though the windings are connected in series to increase the voltage, the current remains the same throughout all the windings.

3. Frogleg armatures are the middle-of-the-road type of windings. The frogleg configuration, the most common of the winding configura-

FIGURE 6–8   Armature
windings.

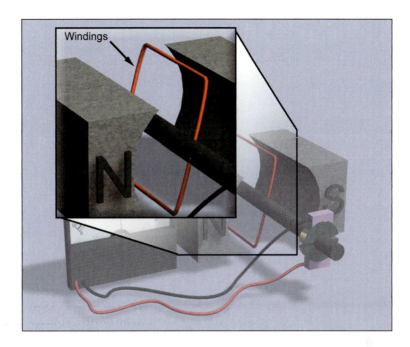

FIGURE 6–9   Types of armature
windings.

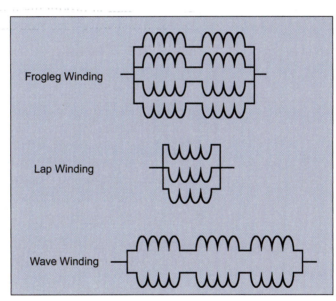

tions, is designed for moderate current and moderate voltage. The frogleg armature, commonly found in larger DC equipment, has its winding connected in series and parallel. That is, there are groups of series-connected windings, and these groups are then connected in parallel.

6.5 Brushes

Brushes ride on the commutator to connect the rotating armature to the load (see Figure 6–10). The brushes are normally made from a carbon-based material. Carbon is used for three reasons:

1. Carbon is softer than copper and allows the brushes rather than the commutator to wear (the latter being more difficult to repair).

FIGURE 6–10   Brushes.

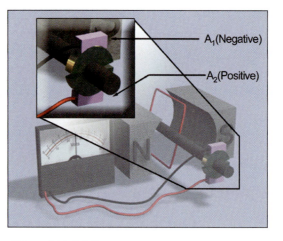

2. Carbon withstands the high temperatures that may be present because of arcing and friction.

3. Carbon, copper, and moisture form a very thin, conductive film on the commutator that lubricates and reduces wear.

The brush leads are identified as A_1 and A_2.

FIGURE 6–11   Housing with pole pieces.

6.6 Pole Pieces

Mounted inside the housing of the generator are pieces of metal, usually made of soft, laminated iron or some other easily magnetized metal. These are fixed pieces and do not rotate. The purpose of the pole pieces is to concentrate the magnetic **field**. The field windings are wound around these pole pieces, forming an electromagnet. The electromagnet sets up the magnetic field necessary for generation. In small generators, the pole pieces are sometimes permanent magnets (see Figure 6–11).

6.7 Field Windings

There are two types of field windings (see Figure 6–12): the series field and the shunt field. Series field windings are connected in series with

FIGURE 6–12   DC generator field windings.

the armature through the brushes. Series windings consist of very few turns and very large wire. The terminal leads of the series are labeled S_1 and S_2.

The shunt field is connected in parallel (shunt) with the armature. They are made with many turns of a very small wire and therefore have a higher resistance than a series field. The shunt field is also called the *field* and is labeled F_1 and F_2.

■ GENERATOR CONNECTION TYPES

6.8 Series Generators

The series generator contains only series field windings connected in series with the armature. The series generator must be self-excited, which means that the pole pieces must contain some amount of residual magnetism to start generating. The residual magnetism produces an initial output voltage that permits current to flow through the field if a load is connected to the generator. In a series generator, the output voltage increases as the load current increases.

As additional loads are connected to a series generator, the output voltage will continue to increase until the magnetic cores saturate (see Figure 6–13). The amount of voltage generated by a series generator is dependent on the strength of the magnetic field produced by the pole pieces, the number of turns of wire on the armature, and the speed of rotation of the armature. Series generators are best suited for use with a constant load.

6.9 Shunt Generators

Shunt generators, the second winding configuration for DC generators, get their name from the field windings being wired in parallel with or shunted across the generator output. Self-excited shunt generators make use of residual magnetism to start the generation process. This is

FIGURE 6–13   Effect of added load on a series DC generator.

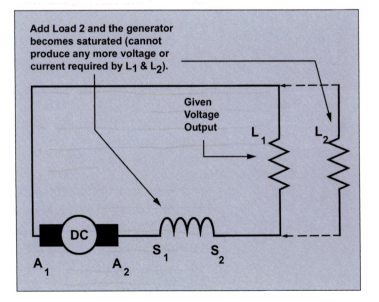

FIGURE 6–14   Shunt field with voltage regulation.

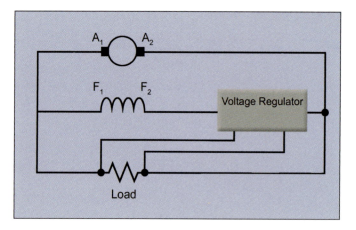

done by applying the voltage generated by the residual magnetic field to the shunt field to create an initial current. The shunt generator provides maximum output voltage before the load is applied. As the load increases on a shunt generator, the output voltage will decrease.

The design parameters of the shunt generator tend to make it self-regulating. As the output voltage drops because of loading or if the output voltage should be shorted, the change in the output voltage will cause a drop to near zero in the field current. With no field current, the generator will shut down.

To attempt to control the output voltage of a shunt generator, an electronic voltage regulator is sometimes installed as part of the generator circuit. The electronic voltage regulator senses the changes in the output voltage and makes adjustments to the shunt field current. The regulator causes more or less current to flow through the shunt circuit when the voltage is sensed to be low or high. Figure 6–14 shows how the voltage regulator, the shunt field, and the output voltage are all related. A simplified explanation of the voltage regulator box is that it acts like a series variable resistor (or rheostat) and voltage divider to the shunt field. This will regulate the output voltage available to the load. In reality, the device is more than a series resistor. The circuitry is much more complex in that it has a controller (sometimes a microprocessor-based computer).

6.10 Compound Generators

Compound generators contain both series and shunt fields (see Figure 6–15). The relationship of the strengths of the two fields in a generator determines the amount of compounding for the machine. Since the voltage regulation of a series generator can be very poor and a shunt generator has good voltage regulation, combining the two gives better circuit flexibility and load control.

There are two subdivisions of the compound generator: short-shunt and long-shunt compounds. The long-shunt compound variation has the shunt winding in parallel with the combined armature and series winding circuit (see Figure 6–15).

The short-shunt configuration has the shunt field connected in parallel with the armature. The series field is then wired in series with the armature/shunt field group (see Figure 6–16).

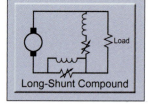

FIGURE 6–15   Long-shunt compound-wound generator.

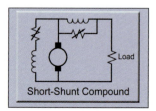

FIGURE 6–16   Short-shunt compound-wound generator.

FIGURE 6–17 Types of generator compounding characteristics.

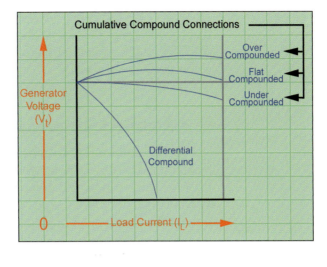

Depending on the number of turns in the series field versus the shunt field, there are three possible ways (classifications) of compounding windings in a generator: overcompounding, undercompounding, and flat compounding. Figure 6–17 shows the typical voltage and current relationships in the different compounds. When the compounding of series and shunt fields aid each other in producing electromagnetic force (inducing voltage and current), the compound is called *cumulative compounding*. When the series windings oppose the shunt field windings in producing electromagnetic force, the compound is called *differential compounding*.

Refer to Figure 6–17. The following are characteristics of the different compound generators:

1. The overcompound generator's full-load voltage is greater than the no-load voltage.
2. The undercompound generator's full-load voltage is less than the no-load voltage, but the series component of the compound makes it a better voltage regulator for loads than the shunt.
3. The flat compound generator has a load voltage equal to the no-load voltage.
4. The differential compound generator output drops to zero in the presence of short circuit. Differentially compounded generators were used for DC welding purposes.

Most commercial compound DC generators (whether used as generators or motors) are normally supplied by the manufacturer as over-compound machines.

■ PARALLELING GENERATORS

There may be a need for connecting DC generators in parallel to supply enough current for a connected load. An equalizer connection is required to prevent one generator from taking the load and the other from acting as a DC motor. Look at Figure 6–18. The equalizing connection is used to connect the series fields of the two machines in parallel with each other. This parallel connection maintains the same voltage across

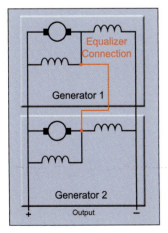

FIGURE 6–18 Parallel generators with equalizing connection.

all series fields and prevents one machine from taking over the other machine as a motor. Notice that the connection must be on the armature side of the series field and connected to the same polarity on each generator.

GENERATOR LOSSES

The power equation for a generator or motor is $P_{in} = P_{out} + P_{loss}$. This means that the power delivered to a generator must always be greater than the output power delivered by the generator to perform work. It also means that not all the power a generator receives can be converted into useful mechanical or electrical energy; there is a loss, and this loss is almost entirely heat. The relationship between power out and power in is called the generator's or motor's *efficiency* (η). The equation for efficiency of a machine is

$$\eta = \frac{P_{out}}{P_{in}} \qquad (6.1)$$

For a generator, Equation 1 becomes

$$\eta = \frac{P_{out}}{P_{out} + P_{in}} \qquad (6.2)$$

and for a motor

$$\eta = \frac{P_{in} - P_{loss}}{P_{in}} \qquad (6.3)$$

Generator power losses can be broken into two main classes: electrical and rotational. Electrical losses are caused by current flow through various parts of the generator and are sometimes referred to as *copper losses*. The greatest of these losses is due to the resistance of the armature. The power loss due to induced currents in the core material of a generator is called *eddy current loss*. This is induced as the armature spins through the flux lines of the pole pieces.

As the DC generator rotates, an alternating current is set up in the armature. This causes the molecules of the iron to realign each time the current changes direction from negative to positive. This molecular friction produces losses called *hysteresis*.

ARMATURE REACTION

In a generator, there are two primary magnetic flux fields: the flux field produced by the field windings around the N and S poles, and the armature flux field. Look at Figure 6–19. Notice that if the armature's magnetic field (flux) shifts the magnetic neutral to maintain a parallel position, the magnetic flux field for the field windings becomes distorted, and a heavy circulating current can be produced that will cause arcing between the commutator segments and the brushes.

FIGURE 6–19 Armature reaction.

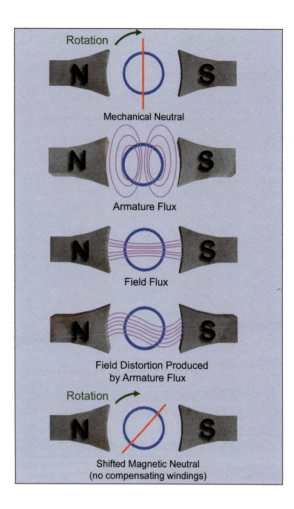

FIGURE 6–20 Neutral plane (with compensating windings).

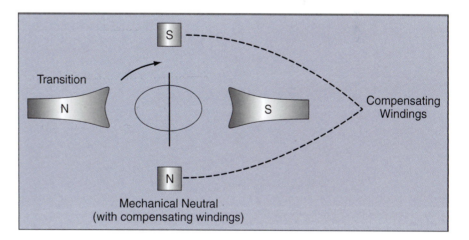

To offset the effect of the distortion from the armature's shift from the magnetic neutral plane (caused by self-induction), compensating windings are used to produce a counterelectromotive force (see Figure 6–20). These windings are placed between the main field poles and are called **compensating**, pole-face, **interpole**, or **Thomson-Ryan windings**.

■ SUMMARY

DC current has some applications that make it superior to AC. For example, DC can provide more precise control. Many industrial plants use DC generators to produce the power needed to operate large DC motors.

Although most generators are AC, there are still some applications for DC. AC generators use slip rings to remove the voltage from the armature; DC generators use commutator segments. The commutator and brushes change the AC produced in the armature into DC. The brushes are used to make contact with the commutator and carry the output current to the outside circuit.

Generator markings include the following:

1. The armature connections are marked A_1 and A_2.
2. The series field windings are marked S_1 and S_2.
3. The shunt field windings are marked F_1 and F_2.

The three factors determining the output of the generator are the following:

1. The number of turns of wire in the armature
2. The strength of the magnetic field
3. The speed of the armature

■ REVIEW QUESTIONS

1. Define a generator.
2. What type of voltage is produced in all rotating armatures?
3. In a DC generator, AC is converted to DC by the _____ and _____.
4. Name three types of armature windings and discuss how they are used.
5. Discuss the three things that determine the voltage output of a DC generator.

6. Discuss the voltage regulation characteristics of the three generator connections.
7. What advantages does a compound-wound generator have over the shunt connection? The series connection?
8. Discuss armature reaction.
    a. What causes it?
    b. How is it compensated?

chapter 7

AC Generators

■ OUTLINE

■ OVERVIEW

English physicist Michael Faraday discovered the basic theory of electrical generation in 1831. He discovered that a magnetic field can be used to produce an electrical current. Practically all commercial power is produced using this principle.

Mechanical energy delivered from raw manpower, water, wind, fossil fuels (coal and oil), and nuclear reactors is used to turn the rotors of a generator. Some of these energy sources, such as fossil fuels and nuclear reactors, must first be converted into another form in order to turn the rotor of a generator.

This chapter covers AC generation fundamentals and shows the relationship between magnetism (which you have previously studied) and the production of electrical power.

OBJECTIVES

After completing this chapter, you should be able to:

1. Describe the operation of AC generators.
2. Identify key parts of the AC generators and their functions.
3. Mathematically determine the relationship between revolutions per minute, frequency, and the number of poles.

■ GLOSSARY

Excitation current   The current supplied to the field of an AC generator. The excitation current creates the magnetic field that the armature cuts through.

PMG   Abbreviation for "permanent magnet generator." In this type of generator, the field is provided by a very strong permanent magnet.

Revolving armature generator   A generator that has the field windings on the stator and the armature windings on the rotor.

Revolving field generator   A generator that has the field windings on the rotor and the armature windings on the stator.

Rotor   The rotating part of the generator.

Round rotor   A more streamlined type of construction where the coils are wound longitudinally on the rotor. This type of construction is generally much lighter and is used extensively on high-speed generators. Also called a *turbo rotor*.

Salient   Projecting or jutting beyond a line or surface; protruding.[1]

Salient pole rotor   A type of rotating field construction where the field poles are wound individually and mounted along the outside edge of the rotor.

Slip rings   Continuous bands of metal installed around a shaft. Slip rings are connected to the rotating windings and provide a path for the current to reach the brushes.

Stator   The stationary part of the generator.

Synchroscope   A type of instrument that is used to determine the phase angle between two voltages. Often used for synchronizing two generators prior to connecting them together.

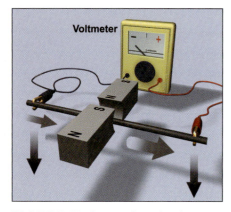

FIGURE 7–1   Induced voltage causing a current flow.

■ INDUCTION BY MAGNETISM

A voltage is induced into a conductor when it is passed through a magnetic field. If the conductor is connected to a complete circuit, current will flow. Figure 7–1 shows a conductor being moved downward between the poles of the two magnets. The poles of the magnets create a field that is moving from north to south. The downward motion of the conductor induces a voltage into the conductor, the polarity of which causes the current to flow and the meter to deflect in the direction shown. The electrons move because of the interactions between the magnetic field of the permanent magnets and the electromagnetic fields of the electrons themselves. Note that without movement, there is no current flow.

If the motion in Figure 7–1 were reversed, it would change the voltage polarity and reverse the flow of current. This is because the direction of the field interactions is opposite. The same would be true if the poles of the magnets were reversed and the motion were downward. The field interactions would be reversed, and current would flow in the opposite direction.

Remember that relative motion between the magnetic field and the conductor must exist before a voltage will be generated. This motion can be the motion of the magnetic field instead of the conductor; that is, the magnets can move instead of the conductor. Movement of the magnetic field will be an important variation in your study of AC generators.

■ GENERATOR PRINCIPLES OF OPERATION

Look at Figures 7–2 and 7–3. This is a repeat of Chapter 6 in the sense that the generating principle is the same. This simple AC generator is constructed of a permanent magnet, **slip rings** (instead of a commutator), brushes, a single loop (armature), and a voltmeter to read generated voltage. The armature is the rotating part of the machine in which the voltage is induced. The magnets set up the required magnetic flux

FIGURE 7–2   AC voltage generator.

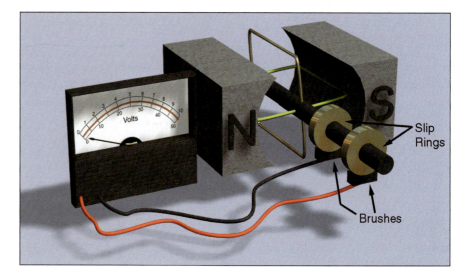

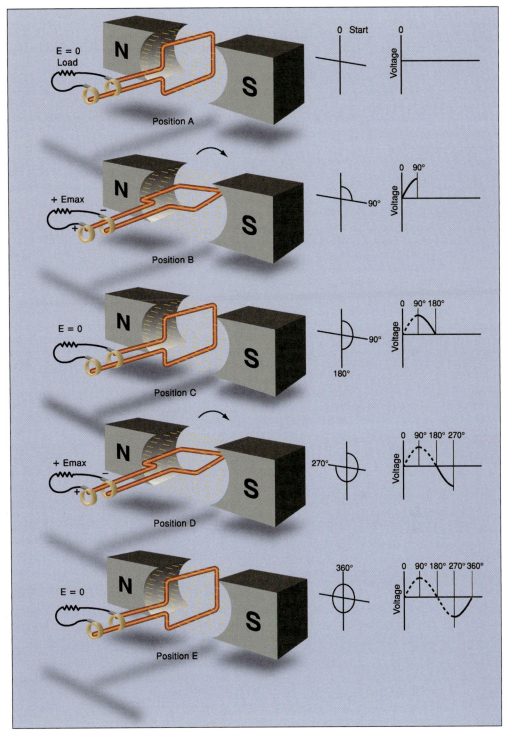

FIGURE 7-3 One cycle of induced voltage.

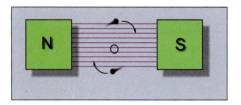

FIGURE 7–4   Cross-sectional view of an AC generator (positions A and E).

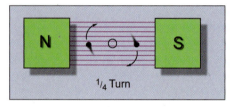

FIGURE 7–5   Cross-sectional view of an AC generator (position B).

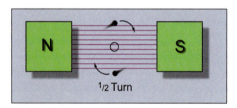

FIGURE 7–6   Cross-sectional view of an AC generator (position C).

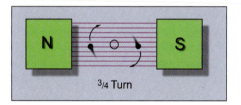

FIGURE 7–7   Cross-sectional view of an AC generator (position D).

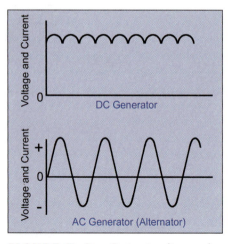

FIGURE 7–8   Output voltage of AC and DC generators.

field. Each end of the wire loop is connected to a slip ring. As the loop rotates, each side of the loop will be cutting lines of flux at the same time. As one side of the loop moves downward, the other side moves upward.

In Figure 7–4 (position A in Figure 7–3), the loop is rotating, but it is cutting very few lines of flux. This position is called the 0° position. If the voltage were being measured here, it would be zero (see Figure 7–3, 0 V).

In Figure 7–5 (position B in Figure 7–3), the generator has rotated 90°, and both sides of the loop are cutting through the lines of flux perpendicularly. At this point, the generator is producing the highest voltage. If the voltage were being measured, it would show the output of the generator to be at its highest "positive" value (see Figure 7–3, +10 V). Notice also that the output increases gradually as the loop rotates from 0° to 90°.

In Figure 7–6 (position C in Figure 7–3), the generator is producing the least amount of voltage again, but this time the loop is in the 180° position, opposite the position shown in Figure 7–5. The zero voltage position is called the *neutral plane* (see Figure 7–3, 0 V).

Figure 7–7 (position D in Figure 7–3) shows the loop in the 270° position. It is now cutting the maximum number of flux lines and producing the maximum voltage. The voltage is now "negative" when compared to the voltage being produced in Figure 7–5. This is because the current is now in the opposite direction through the loop (see Figure 7–3, −10 V).

■ CONSTRUCTION OF AC GENERATORS

AC generators, also known as alternators, operate on the same principle of electromagnetic induction as DC generators. Both types of generators pass conductors through a magnetic field to create a voltage and resultant current flow. However, unlike the DC generator, which uses a commutator to rectify the generated signal, the AC generator uses slip rings to connect the AC output directly to the load. The different outputs of DC and AC generators can be seen in Figure 7–8.

There are two basic types of alternators: **revolving armature** and **revolving field**. The revolving armature is used primarily for small-load applications; however, both revolving armature and revolving field AC generators produce an alternating output that goes both positive and negative.

7.1 Revolving Armature AC Generators

The revolving armature AC generator is not as common as the revolving field generator. It is similar in construction to the DC generator in that both types use some form of mechanical device to connect the external circuit to the armature circuit. The AC generator uses slip rings instead of a commutator. The purpose of the slip rings is to directly connect the armature windings to the load. Since no commutator is used, the output is an alternating voltage instead of a direct voltage.

The slip ring, unlike the segmented DC generator commutator, is constructed as a continuous ring and is not broken up into isolated seg-

FIGURE 7–9   Simple revolving armature generator.

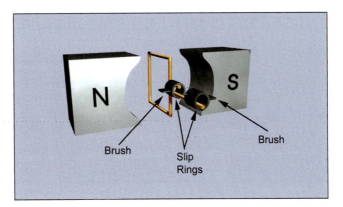

FIGURE 7–9   Simple revolving armature generator.

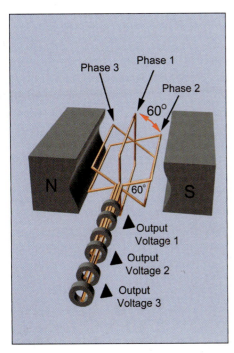

FIGURE 7–10   Three-phase revolving armature AC generator.

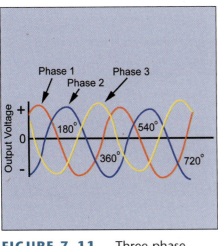

FIGURE 7–11   Three-phase generator output.

ments. Each individual end of the rotating loop of the armature is connected to a separate slip ring (see Figure 7–9).

In a three-phase revolving armature AC generator, there are three separate sets of slip rings, one set for each phase. Each set would be connected to a set of windings, or loops, that are electrically oriented 120° from each other (see Figure 7–10). The output of the three-phase generator would resemble the output wave, as seen in Figure 7–11.

Although the revolving armature AC generator can produce an alternating output voltage, the slip rings and brushes limit the size of load it can carry. The higher currents require large slip rings and large brushes and highly increased maintenance. The high voltages required by some loads mean that much heavier insulation must be used. Taken together, these features limit the kilovolt-ampere (kVA) capacity of the revolving armature generator.

7.2 Revolving Field AC Generators

This type of generator has the armature windings on the **stator** and the field windings on the **rotor**. By mounting the windings that provide the output voltage in a stationary position and rotating the magnetic field, the capacity and connection limitations experienced with the revolving armature type of AC generator are eliminated. Higher voltage and kVA ratings are possible with the revolving field AC generator because the circuit to the outside can be hardwired instead of connected through a brush and slip ring arrangement.

The portion of the generator that rotates is the rotor, and the stationary portion is the stator. The stator portion of the AC generator is made up of sets of windings. Figure 7–12 shows an example of a three-phase revolving field generator. This type of generator stator is constructed by placing the three sets of windings 120° apart.

7.3 Rotor Construction

Rotating fields are constructed in two different ways: the **salient pole rotor** and the **round rotor**. The round rotor is also called a *turbo rotor*. Figures 7–13 and 7–14 are drawings of the salient pole rotor and the round rotor, respectively.

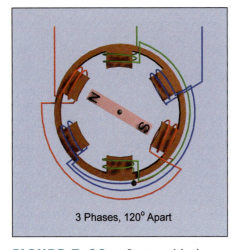

FIGURE 7–12   Stator with three-phase windings.

Salient Pole Construction

Figure 7–13 shows the salient pole construction. This particular rotor is one that is used on a very large hydroelectric generator. Much smaller salient pole rotors are used in some transportable generators.

Notice the pole faces arranged around the outside of the rotor. They jut out and are very pronounced. This is why they are called *salient poles*. Each one of the pole faces will be magnetized by its own set of coils. North and south poles are adjacent to each other so that every north pole is between two south poles and vice versa.

This type of rotor construction is very heavy because of the iron pole faces and the copper windings. Such rotors are subject to very high wind resistance and centrifugal forces.

Round Rotor Construction

As you can see in Figure 7–14, the round rotor has windings that are wound longitudinally. Round rotors are generally smaller and lighter and are found in the high-speed two- or four-pole generators used in large utility generating stations. Although such a rotor usually turns much faster than the salient pole rotor, its longer, narrower construction minimizes wind resistance and centrifugal forces.

7.4 Excitation Systems

Regardless of whether the field is revolving or stationary, it must create a magnetic field so that the armature windings can cut through it (or be cut through, in the case of the revolving field) and produce the output voltage. Many different types of excitation systems are used. A few of the more important ones are discussed here.

Permanent Magnet Fields

Some smaller generators use a permanent magnet to provide the field. This type of construction is usually limited to small auxiliary generators used for excitation or control purposes. Such a generator is called a **PMG** (permanent magnet generator).

FIGURE 7–13   Salient pole rotor (used on large hydroelectric generators).

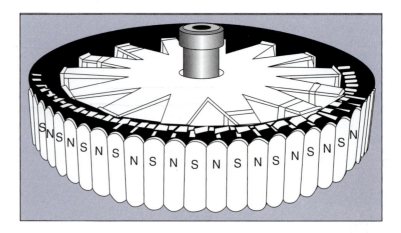

FIGURE 7–14   Round rotor (turbo rotor).

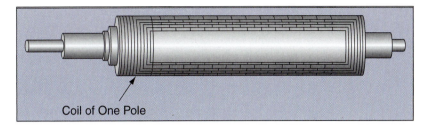

Coil of One Pole

Electromagnetic Field

Generators use some sort of DC supply to provide the current for their excitation. The equipment that provides this current is called an *exciter* and the current it provides is called **excitation current**. Exciters are made in a variety of different ways, including the fixed exciter and the rotating exciter.

An example of a fixed exciter for a revolving field generator is shown in Figure 7–15. Here a battery is connected to the windings of the field magnets mounted on the rotor. The battery supplies the DC excitation that creates the current revolving field that cuts through the armature. The output of the armature is then connected to the load. In this case, the battery is the exciter for the generator.

A block diagram of a fixed exciter is shown in Figure 7–16. In this type of exciter, the battery is replaced with an AC supply, a rectifier, and a DC voltage regulator. The transformer changes the voltage to a level suitable for the exciter input. The regulator controls the generator voltage output by regulating the amount of DC current supplied to the field windings. A sample of the generator output voltage is fed back to the regulator as the reference for voltage control.

A rotating exciter is a separate DC generator that supplies the DC current to the field windings of the main generator. This type of exciter can be separate from the main generator or mounted on the same shaft as the main generator. If it is mounted on the same shaft, it is usually called a *brushless exciter.*

FIGURE 7–15   A battery used to provide excitation current.

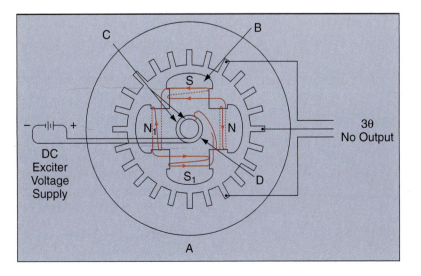

FIGURE 7–16 Static exciter using a transformer and a voltage regulator.

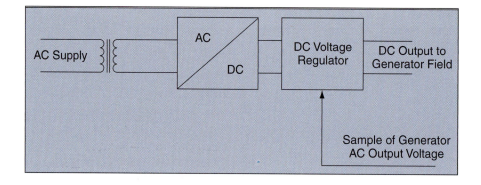

In a brushless exciter, the armature of the exciter is mounted on the same shaft as the field windings of the main generator. Since the two coils are rotating on the same shaft, there is no need for brushes or slip rings. However, since the output of the exciter armature is AC, it must be changed to DC. This is done by attaching a circular disk to the shaft. The disk has a diode rectifier mounted on it. The output from the exciter is connected to the input of the diode rectifier. The output of the diode rectifier is connected directly to the rotating field windings. The voltage magnitude on the system is controlled by regulating the field current of the exciter.

7.5 Generator Cooling

Generators with small kVA outputs are normally air cooled. Appropriate openings are left in the stator windings, and slots are provided for air to pass through. Air-cooled alternators will usually have a fan attached to one end of the rotor shaft that helps circulate the air through the entire unit.

Large-capacity alternators are often enclosed and operate in a hydrogen atmosphere. There are two reasons for using hydrogen as a "cooling" atmosphere. One is that the hydrogen is less dense, which means there is less wind resistance; therefore, less heat is produced. The other is that the hydrogen atmosphere dissipates heat faster than air does.

▪ FREQUENCY AND OUTPUT VOLTAGE

7.6 Frequency Factors

The frequency output of the voltage is a critical aspect of the AC generator's design. The output frequency is determined by the number of stator poles and the speed of rotation of the rotor. Because the number of stator poles is constant and is determined by the design of the generator, the only way to vary output frequency is by the speed of the rotor.

The formula for calculating the frequency when the poles and revolutions per minute (rpm) are known is

$$f = \frac{S \times P}{120}$$

(7.1)

Where:

f = frequency in Hz

P = number of poles

S = speed in rpm

120 = a constant.

The number of poles is the total number of north and south poles present in the generator stator. Since they always come in pairs, the number of poles will always be an even number. Two- and four-pole machines are the most common construction types for generators, but many hydroplants and other such generators may have hundreds of poles.

The 120 number is actually 2 × 60. The 2 changes the number of poles to the number of pole pairs, and the 60 changes rpm to revolutions per second.

Table 7–1 shows the relationship among frequency, speed of rotation, and number of poles for a variety of standard generator types. All the numbers in Table 7–1 were developed using Equation 1.

EXAMPLE 1

What is the output of an alternator that contains four poles and is turning at a speed of 1,500 rpm?

Solution:

$$f = \frac{S \times P}{120} = \frac{1,500 \times 4}{60} = 100 \text{ Hz}$$

7.7 Output Voltage

Output voltage is just as important as frequency. Remember that the output voltage is a function of how many lines of flux are being cut

Table 7–1   Speed–Frequency Relationships in an AC Generator

Number of Poles	Desired Frequency		
	25 Hz	50 Hz	60 Hz
2	1,500 rpm	3,000 rpm	3,600 rpm
4	750	1,500	1,800
6	500	1,000	1,200
8	375	750	900
10	300	600	720
12	250	500	600
14	214.3	428.6	514.3

each second. With that in mind, you can see that there are three generator parameters that will affect the voltage:

1. The length of the armature conductors, determined by the length of each turn and the total number of turns
2. The speed of rotation of the generator
3. The strength of the magnetic field

The number of turns is fixed by the generator's design, and the speed of rotation is fixed by the output frequency requirements. Thus, the only option for voltage control is to control the strength of the magnetic field. Since the excitation current controls the magnetic field's strength, it is used for the generator output voltage control.

■ PARALLELING GENERATORS

There are three conditions that must be met when paralleling alternators:

1. The phase rotation of the two machines must be the same. This means that the A, B, and C phases must occur in the same sequence on both machines.
2. Each phase on one machine must be in phase with the equivalent phase on the other machine (see Figure 7–17).
3. The output voltage of the two machines must be the same.

One of the most common methods of detecting when the direction of the magnetic field or the phase rotation of one alternator is matched with that of a paralleled alternator is to use three lamps. Each lamp is powered by the voltage between the phases of each generator. The Aϕ voltage of generator 1 will connect to one side of a light, and the Aϕ voltage of generator 2 will connect to the other side. When the two generators are in phase and the voltage magnitudes are equal, the lights will be off. This indicates synchronism.

Another instrument that can be used to parallel two machines is called a **synchroscope**. This instrument measures the voltage and frequency differences of two alternators being placed in parallel. The base load alternator (the one previously connected to the load) is used as the

FIGURE 7–17   Generator output voltages when the generators are synchronized.

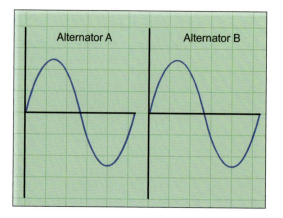

Alternator A   Alternator B

reference point for measuring the frequency and voltage of the second alternator.

The synchroscope meter display will indicate whether there is a speed/frequency mismatch and/or a voltage mismatch between the two machines being paralleled. The electromechanical synchroscope looks somewhat like a clock with only one hand. The two voltages are connected to the inputs, and if they are in phase, the meter hand will point straight up (12 o'clock). The meter can rotate a full 360° in a clockwise direction. When the frequency or phase angle changes, the meter rotates to a new position.

■ SUMMARY

There are two basic types of three-phase alternators: the revolving armature and the revolving field. The revolving armature is least used because of the limited power and voltage rating caused by brush and slip ring limitations.

The rotor of the revolving field alternator has electromagnets. Direct current supplied to the field is called excitation current. The number of poles and the speed of the rotor determine the output frequency.

The output voltage is controlled by the DC excitation current. This increases or decreases the strength of the rotor's magnetic field and results in induction of more or less output voltage and current.

■ REVIEW QUESTIONS

1. Discuss the creation of electricity through electromagnetic induction.

   a. If you have a wire and a magnetic field, how do you create voltage on the wire?

   b. How do you reverse the polarity of the voltage?

   c. How do you increase or decrease the magnitude of the voltage?

2. Using Figure 6–7b, explain how an AC generator creates a sine wave.

3. What is the physical difference between a DC generator and an AC generator with a revolving armature?

4. Discuss the difference between an AC generator with a revolving armature and one with a revolving field.

   a. Which one would normally be used for larger voltages and current?

   b. What sort of maintenance problems would exist for each of them?

5. What is the difference between a salient pole rotor and a turbo rotor?

6. Discuss the relationship between speed of rotation, number of poles, and output frequency.

7. How do you increase or decrease the output voltage of an AC generator without changing its output frequency?

PART
3

INDUCTANCE IN AC CIRCUITS

chapter **8**

Inductance and Its Effects on Circuits

■ OUTLINE

■ OVERVIEW

Inductance and capacitance are major factors in analyzing and understanding AC circuits. This chapter introduces many of the key aspects of inductance and its sources. Two different aspects of inductance are covered: the fundamentals of electromagnetism and the physical aspects of inductance and how they affect electrical properties of the circuit.

The chapter lays the foundation for solving problems in AC circuits and understanding the operation of inductive devices such as transformers. Chapters 9 and 10 present more detail on inductive reactance and will show you how to manipulate inductors during circuit analysis.

■ OBJECTIVES

After completing this chapter, you should be able to:

1. Define inductance and self-inductance.
2. Describe the counterelectromotive force using Lenz's law.
3. Identify the physical factors of the effect of inductance.
4. Explain the behavior of current in DC and AC circuits when inductance is present.
5. Describe the relationship among current, applied voltage, and counterelectromotive force in inductive circuits.

■ GLOSSARY

Back EMF   See Counter EMF.

Counter EMF   The voltage induced in an inductor by the changing magnetic field.

Electromotive force   See Voltage.

Henry   The unit of inductance. A coil has an inductance of 1 henry when a current change of 1 ampere per second causes an induced voltage of 1 volt.

Inductance   The property of an electric circuit by virtue of which a varying current induces an electromotive force in that circuit or in a neighboring circuit.[1]

Leakage flux   Flux lines that do not link properly in an inductor. Leakage flux re-

duces the overall magnetic field and the resulting inductance.

Permeability   The ability of a material to concentrate or focus magnetic flux lines. Absolute permeability is measured in henries per meter. The permeability of a vacuum is 1.26×10^{-6} H/m.

Self-inductance   The property of an electrical component (such as a coil of wire) to induce a voltage into itself as the current through the component changes.

Voltage   The electrical force created between two areas of different electrical charge.

---

[1]Excerpted from the *IEEE Standard Dictionary of Electrical and Electronic Terms* (Std-100. 1998 Fourth Edition).

FIGURE 8–1   Current flow
producing a magnetic field.

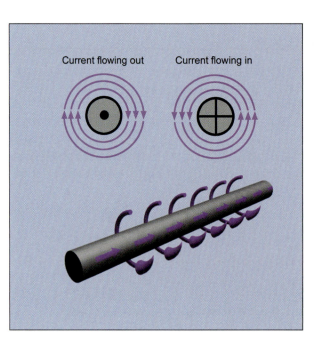

■ MAGNETIC INDUCTION

One of the basic rules of electricity is that current flowing through a conductor produces a magnetic field (called *magnetic flux*) around the conductor. As with many electrical laws and principles, this one also works in reverse; that is, a **voltage** is induced when a conductor cuts through magnetic lines of flux or when magnetic lines of flux cut through a conductor.

The ability of a changing or moving magnetic field to induce a voltage into a conductor is called *electromagnetic induction*. When an electrical current flows in a circuit, a magnetic field is always created (see Figure 8–1).

When no current flows through a conductor, the free electrons move randomly, and their magnetic fields cancel each other out. When a voltage is applied, those free electrons line up and flow in one direction, and their magnetic fields add up.

Figure 8–2 illustrates the key components of electromagnetic induction. When the conductor is moved downward through the magnet's lines of flux, the induced voltage will cause an **electromotive force** (EMF) on the electrons. If there is a circuit, as there is in Figure 8–2, the EMF will cause a current flow in one direction. There must be relative motion between the conductor and the magnetic field. If the direction of relative motion is changed, current flow will change direction. Cutting more lines of force per second will increase the amount of EMF and the resulting current. From this fact, you can see that moving the wire perpendicular to the lines of force at a given speed will create more voltage than if the wire is moved at some angle other than 90°.

If the conductor in Figure 8–2 moves upward instead of downward, the polarity of the induced voltage and direction of current flow will be reversed. Therefore, the polarity of the induced voltage is determined by the polarity of the magnetic field relative to the movement of the conductor.

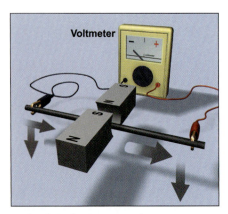

FIGURE 8–2   Conductor cutting
magnetic flux to produce voltage.

What about moving the magnetic field? Remember that the important factors concerning magnetic induction are a conductor, a magnetic field, and movement between the two. In practice, it is often preferable to move the magnet instead of the conductor. Previously you learned that most AC generators operate on this principle. In Figure 8–2, moving the magnets instead of the conductor will produce the same result as long as the speed and direction of relative motion are the same.

■ INDUCTANCE

Inductance is the ability of a current-carrying conductor, coil, or circuit to induce a voltage into itself or adjacent circuits. **Self-inductance** is the ability to induce a voltage into itself. There is some self-inductance present in all AC circuits because of the continuously changing magnetic fields.

Consider the coil shown in Figure 8–3. The magnetic field created by the changing current through the conductor expands and collapses with the current. This changing magnetic field cuts through the turns of the coil, inducing a voltage in the coil that opposes the change in current. All loads, such as motors and transformers, contain these coils.

The total effect of all this can be summarized in three statements:

1. As the current increases, the magnetic field that it causes builds (increases).

2. As the current decreases, the magnetic field that it causes collapses (decreases).

FIGURE 8–3 Current flow and the resulting magnetic field through a coil.

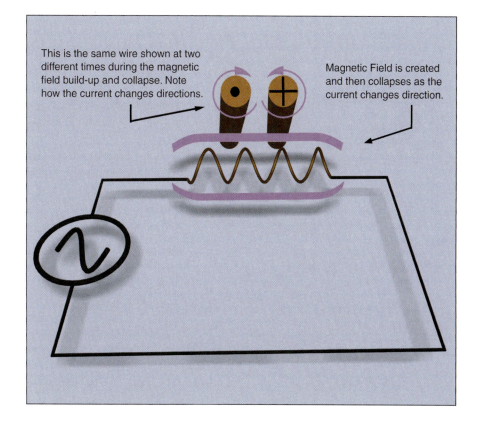

This is the same wire shown at two different times during the magnetic field build-up and collapse. Note how the current changes directions.

Magnetic Field is created and then collapses as the current changes direction.

Table 8–1   Magnetic Field Action in an Inductor

Applied Voltage	Current Flow	Magnetic Field Created by Current	Voltage Induced by Magnetic Field Change
Constant (DC)	Constant	Constant	Zero
Decreases	Starts to decrease	Decreases	Opposes change
Increases	Starts to increase	Increases	Opposes change

3. When the current changes direction, the magnetic field it creates changes polarity.

Table 8–1 illustrates the entire process and introduces some very important considerations.

This table shows that as the magnetic field collapses, it creates a voltage that opposes the original change in voltage. In other words, if the current decreases, the collapsing magnetic field cuts through the coil conductors and induces a voltage that tries to prevent the decrease. If the current increases, the expanding magnetic field cuts through the coil conductors and induces a voltage that tries to prevent the increase. The magnitude of the opposing voltage created will be determined by how fast the current is changing. Remember that the voltage induced is a function of how many lines of flux are being cut each second. Thus, this overall effect is one in which the increasing or collapsing magnetic field induces a voltage that is 180° out of phase with the applied voltage.

For example, assume an inductor with no resistance connected to a 120-VAC supply. Further assume a point in time when the collapsing (or increasing) magnetic field is inducing a voltage of 108 V into the coil. Since an equal amount of applied voltage must be used to overcome the induced voltage, there will only be 12 V to push the current through the conductor. Figure 8–4 shows the 180° phase relationship between the applied voltage and the induced voltage as well as the vector calculations for the resultant voltage.

Since the induced EMF appears only during a change in the current flow, the magnitude of the self-induced EMF depends on the rate of change of the current and the amount of the inductance. The induced EMF is also called the **counter EMF** and is abbreviated V_{CEMF}. Later you will learn that

$$V_{\text{CEMF}} = -L \times \frac{\Delta I}{\Delta t} \tag{8.1}$$

8.1 Lenz's Law

A coil of wire wrapped around a core is called an *inductor*. Its name comes from the fact that a voltage can be induced into it by moving it through a magnetic field or by moving a magnetic field around it. This induced voltage produces a current. You also learned that passing current though a wire produces a magnetic field around it.

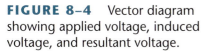

FIGURE 8–4 Vector diagram showing applied voltage, induced voltage, and resultant voltage.

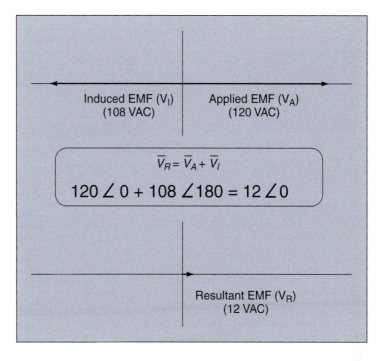

Heinrich Lenz, a German physicist, discovered that the magnetic field that induced the voltage and the magnetic field that was produced by the induced current oppose each other. This is the basic principle used to determine the direction of an induced voltage. Lenz's law states that an induced voltage or current opposes the motion that created the induced voltage or current. The induced voltage and related current is 180° out of phase from the applied voltage and resulting current.

When the current through an inductor changes, the magnetic field around the inductor will change. This change in magnitude causes the field to move (expand or collapse) relative to the coil. The relative motion will induce a voltage that opposes the change in current that produced the change in the field. Figure 8–5a shows the phase relationship between the applied voltage and the induced voltage. The basic rule is that inductors always oppose a change in current.

It is important to understand the concept of applied voltage (EMF) and induced voltage (counter EMF) and their respective currents as they relate to the magnetic field surrounding a conductor. Figure 8–5b may help explain that relationship. The following explanations reference the sine wave shown in Figure 8–5b and the numbered points along that waveform.

1. With no current flowing, there is no field.
2. As the current flow begins to increase, the field begins to expand, cutting the conductor in the direction shown (moving outward).
3. As the current flow continues to increase, the field continues to expand.
4. The current is at its maximum and is momentarily unchanging. The field is also not changing (not moving) because of the stable state of the current flow.

FIGURE 8–5   (a) Induced and applied voltage are 180° out of phase; (b) magnetic field of an inductor with an alternating current.

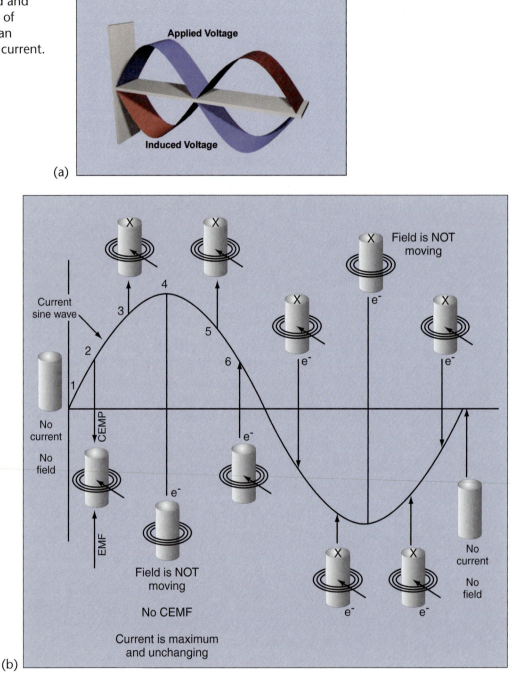

(a)

(b)

5. After having been at the maximum, the current flow begins to decrease (change), and the magnetic field begins to react by collapsing inward (changing the direction in which it is moving relative to the conductor). The field does not change polarity, just strength.

6. The current continues to decrease, and the field continues to decrease relative to the current flow.

7. The current will change direction and is increasing but this time in a negative direction. Since the current is increasing in the opposite direction, the magnetic field expands outward but with reverse polarity (polarity opposite to what was found during the positive portion of the sine wave).

8. As the current flow continues to expand, the field also continues to expand.

9. The field is at its maximum as a result of the current being at its maximum negative value. The magnetic field is neither expanding nor decreasing at this point.

10. After having been at the maximum negative value, the current value begins to decrease (become less negative), and the field will also continue to decrease (collapse inward, changing the direction in which it is moving relative to the conductor); however, the polarity will not change.

11. As the current continues to decrease, the field will also continue to decrease.

12. With no current flow, there is no field.

■ ELECTRICAL AND PHYSICAL PROPERTIES OF INDUCTANCE

8.2 Counter EMF

The voltage created by the magnetic field that is 180° out of phase with the original voltage is called *counter EMF* because it opposes, or *counters*, the applied EMF. Another term for counter EMF (CEMF) is **back EMF**. This whole concept is shown in Figure 8–6. CEMF opposes current when it is increasing or decreasing.

8.3 The Henry

Inductance is the property of a conductor that opposes any change in current and is measured in units called the **henry** (H). In calculations, the inductance is represented by the letter "L." A coil has an inductance of 1 henry when a current change of 1 ampere per second causes an induced voltage of 1 volt. Knowing that the inductance is the quantity that causes a counter EMF when the current changes leads to the formula that ties these three quantities together:

$$V_{\text{CEMF}} = -L \times \frac{\Delta I}{\Delta t} \qquad (8.2)$$

FIGURE 8–6 Applied and induced voltage opposing each other.

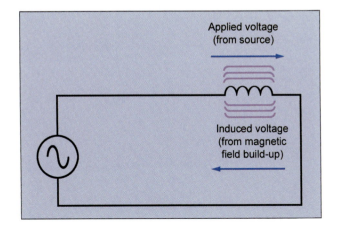

Applied voltage
(from source)

Induced voltage
(from magnetic
field build-up)

The minus sign indicates that the voltage is in opposition to the changing current that is causing it. The Greek letter delta (Δ) stands for "change in," so ΔI means the change in current and Δt the change in time. This is commonly spoken as the "delta I" or "delta t." Given the CEMF, a current change, and the time differential, you can calculate the inductance using the formula in Equation 2.

EXAMPLE 1

A current change in a certain coil is 2 amps in 0.1 second. This change causes an induced voltage of 10 volts in the coil. Calculate the inductance.

Solution:

$$V_{CEMF} = -L \times \frac{\Delta I}{\Delta t} = -L \times \frac{2}{0.1} = -10$$

Solving for L gives

$$L = -\frac{0.1}{2} \times -10 = 0.5 \text{ H}$$

8.4 Physical Factors Affecting Inductance

There are a number of factors that determine the amount of inductance in a coil. The following paragraphs describe those factors.

Number of Turns of Wire

The self-inductance of a wire or arrangement of wires is directly related to the number of times the wire is coiled. To understand why this is, start by looking at Figure 8–7. In the figure, a conductor has been looped into one coil. Using the left-hand rule for the magnetic field, you can see that the fields from each side of the coil add to the fields

FIGURE 8–7   Magnetic flux around a single coil.

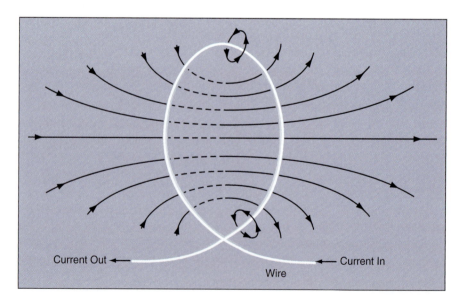

Current Out ←

Current In →

Wire

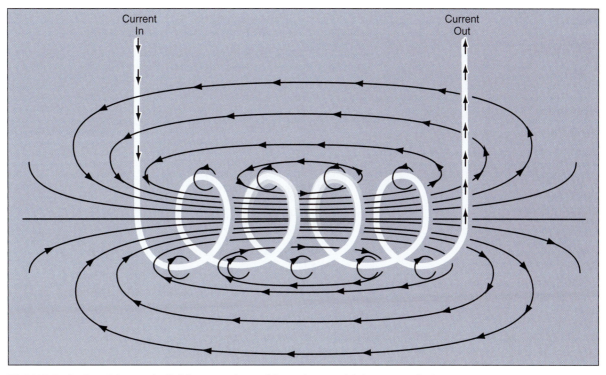

FIGURE 8–8   Magnetic field strengthened by using multiple turns of wire.

set up by their opposite sides. Therefore, by creating a single coil, the number of flux lines and hence the inductance have been doubled.

Figure 8–8 shows what happens when more turns are added. The physical arrangement causes each individual coil's field to add to those of the other coils. Clearly, the total number of flux lines and the inductance are directly related to the number of turns of wire.

Spacing between the Turns

Notice in Figure 8–8 that not all of each coil's flux lines link all the other coils. Some of the flux lines from one coil may link with only one other coil, two other coils, or perhaps none at all. The flux lines that do not link all the other coils do not contribute to total inductance and are called **leakage flux**. One way to reduce the amount of leakage flux is to wind each individual coil closer to the others.

Cross-Sectional Area of the Core

The inductance of a coil is also related to the density, or concentration, of the flux lines. To understand this, think of the magnetic field of the earth. The intensity of the earth's magnetic field is immense, yet it does not "snatch the watch right off of your wrist." The reason is that the earth's lines of flux are spread across a vast area. The total strength of any magnetic field is a function of both the number of lines and how dense they are.

Permeability of the Core Material

Magnetic **permeability** is the ability of a material to focus, or concentrate, magnetic lines of flux. Iron has a permeability that is on the order of 1,000 times greater than air. This means that if an iron core is

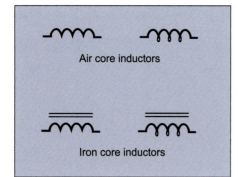

FIGURE 8–9   Schematic symbols for inductors.

inserted into a coil, the lines will be concentrated, and the resulting magnetic field and inductance will be increased.

When all the effects of these various characteristics are evaluated and measured, the following formula can be generated for a coil with a circular core:

$$L = \frac{0.4\pi N^2 \, \mu A}{l} \qquad (8.3)$$

Where:

L = inductance in henrys
N = number of turns
μ = permeability of the core material
A = cross-sectional area of core in square meters
l = length of core in meters
0.4π is a constant for a circular core inductor.

For other shaped cores (oval, square, or triangular), different equations apply.

Inductance has no effect on a DC circuit as long as the current is constant. It does have an effect if the supply is turned off and on or when the DC changes in value.

8.5 Schematic Symbols

The schematic symbol for an inductor looks like a coil of wire, as shown in Figure 8–9. The symbols shown with the two parallel lines represent iron core inductors. The symbols without the parallel lines represent air core inductors.

■ CIRCUIT INDUCTANCE

8.6 Rise Time of Inductor Current

When DC voltage is applied to a resistive load, as shown in Figure 8–10, the current will instantly rise to its maximum value. A resistor has a value of 6 Ω, and it is connected to a 12-V source. When the switch is closed, the current will rise instantly to a value of 2 amps: $\frac{12\text{ V}}{6\,\Omega} = 2$ A.

If an inductor is added to the circuit, as shown in Figure 8–11, and the switch is closed, the current cannot change instantaneously because of the CEMF produced by the inductor. When the switch is first closed, the current tries to rise to its maximum value, just as it did in the circuit in Figure 8–10; however, the sudden change in current (from zero) will cause the inductor to generate a CEMF to resist that attempt to immediately rise to the maximum value. Remember that $V_{\text{CEMF}} = -L \times \frac{\Delta I}{\Delta t}$.

The CEMF opposes the current and increases the amount of time required for it to reach maximum. As the current flow increases and

FIGURE 8–10 DC current rise in a resistive circuit.

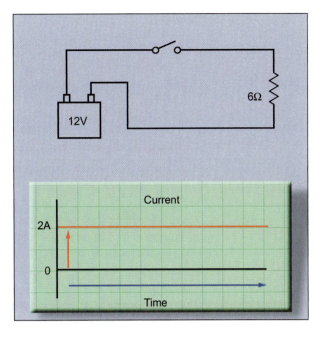

FIGURE 8–11 DC current rise in an inductive circuit.

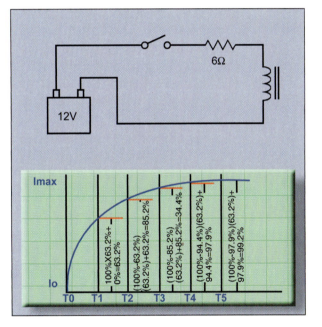

approaches the maximum current flow value, the rate of change decreases, as does the amount of induced voltage. When the current reaches (or gets very close to) its maximum, the inductor no longer creates any CEMF; consequently, the maximum current in both circuits (Figures 8–10 and 8–11) is the same.

The amount of time required for the current flow to reach the maximum value is expressed in time constants. In a circuit with inductance and reactance, the time constant is defined from the following formula:

$$T = \frac{L}{R}$$

(8.4)

Where:

T = the time constant in seconds
L = the circuit inductance in henrys
R = the circuit resistance in ohms.

As the switch is initially closed, the circuit experiences the greatest amount of change in current flow. The circuit reaches 63.2% of the maximum current flow in the first time constant. As the circuit proceeds through the next four time constants, the amount of change gets smaller with each time constant and gradually tapers off as the current reaches the maximum value. The amount of change in current from 0% to 100% and the number of time constants (five) necessary to reach the maximum constant current will be the same regardless of the size of the inductor and the source voltage. The total time in seconds varies depending on the circuit values because the time constant in seconds varies. This type of current change is called an *exponential change*.

The inductance works during charge and discharge. Just as the circuit takes five time constants to reach a value very close to the maximum circuit current when the switch is closed, the same circuit will also take five time constants to reach a near-zero value once the switch is opened. Look at Figure 8–12. Assume that switch S_1 is closed and the current flow has reached its maximum value of 2 amperes. If S_1 is suddenly opened and S_2 is simultaneously closed, the inductor magnetic field will start to collapse. As it does, the CEMF created in the inductor will cause current to flow through the resistor and switch S_2. Initially, the current will be 2 amperes since the current through an inductor will not change instantaneously. Gradually, however, the current will decay, and it will do so in the same amount of time it took to rise to the final value.

8.7 Exponential Rates and Time Constants

The exponential curve describes a rate of occurrences in nature. The curve is divided into five time constants or divisions. Each time constant will have a change equal to 63.2% of the remaining difference or value. During the first time constant, the value will rise from 0% to 63.2% of its total value, during the second time constant it will rise to 63.2% of what is left to the maximum, and so on.

FIGURE 8–12 DC current decay in an inductive circuit.

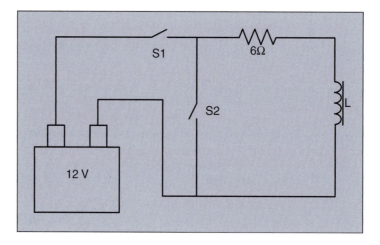

EXAMPLE 2

It takes 5 seconds for the current to rise to a maximum 10 amps in a circuit. This means that the one time constant is 1 second. Table 8–2 shows the rise times and the total voltage for each time constant.

Two important points need to be made about this type of current (or voltage change):

1. Theoretically, the current will never reach the final value. You can see this by inspecting the values in Table 8–2. Notice that with each time constant, the total current gets closer to the final value. Notice also, however, that with each time constant, the change in current gets smaller. In this example, the current actually reaches only 99.3% of the maximum. Another time constant would make it closer but still not reach maximum.

2. The increasing waveform, as shown in Figure 8–11, can be expressed mathematically as

$$I = I_{\max} \times \left(1 - e^{\frac{-t}{T}} \right) \qquad (8.5)$$

where T is the time constant and $I_{\max}$ is the maximum current value.

8.8 Inductive Reactance

Just as resistance offers opposition to the current flow in a DC circuit, an inductor offers opposition to the current flow in an AC circuit. The opposition to AC by an inductor is called the *inductive reactance* and is given the symbol X_L. The key elements that affect the magnitude of inductive reactance are the frequency and the inductance of the coil. The inductive reactance in a circuit (X_L) can be calculated by the following formula:

$$X_L = 2\pi f L \qquad (8.6)$$

Where:

X_L = the inductive reactance in ohms
f = frequency in hertz
L = inductance in henrys.

Table 8–2   Solution to Example 8–2

Time	Rise	Final Voltage
1	63.2% × 10 amperes = 6.32 amperes	6.32 A
2	63.2% × (10 A − 6.32 A) = 2.33 A	2.33 A + 6.32 A = 8.65 A
3	63.2% × (10 A − 8.65 A) = 0.853 A	0.853 A + 8.65 A = 9.5 A
4	63.2% × (10 A − 9.5 A) = 0.316 A	0.316 A + 9.5 A = 9.81 A
5	63.2% × (10 A − 9.81 A) = 0.12 A	0.12 A + 9.81 A = 9.93 A

As can be seen from the formula, an increase in the frequency of the circuit and/or an increase in the size of the inductor will cause a higher opposition to current flow. In the theoretical circuit that has only inductive reactance opposing the flow of current in the circuit (assuming that the circuit conductors and the conductor material that makes up the inductor itself have no resistive value), the inductive reactance value can be directly substituted for the resistance value in the Ohm's law equation to calculate the current flow. That formula is

$$E = I \times X_L$$

or

$$I = \frac{E}{X_L} \tag{8.7}$$

or

$$X_L = \frac{E}{I}$$

As can be seen in the Ohm's law relationships in these equations, changing the frequency, which will change the X_L, will change the circuit current.

EXAMPLE 3

A 10-H inductor is placed in series with a 100-VAC source with a frequency of 60 Hz. What is the inductive reactance of the circuit? How much current will flow?

Solution:

$$X_L = 2\pi f L = 2\pi \times 60 \text{ Hz} \times 10 \text{ H} = 3{,}770 \ \Omega$$

$$I = \frac{E}{X_L} = \frac{100 \text{ V}}{3{,}770 \ \Omega} = 26.5 \text{ mA}$$

Remember that frequency is the variable in this equation. The calculated X_L and I are for a given frequency and will change if the frequency changes.

8.9 Series Inductors

When inductors are connected in series, the total inductance is the sum of the inductance of all the inductors. Note that this is the same as for resistance in series for Ohm's law in equation form:

$$L_{tot} = L_1 + L_2 + L_3 + \ldots L_N \tag{8.8}$$

where L_{tot} equals the total circuit inductance, L_1 equals the value of inductor 1, L_2 equals the value of inductor 2, L_3 equals the value of inductor 3, and L_N equals the value of the Nth inductor. This is somewhat intuitive if you imagine that instead of connecting the inductors in series, you are just adding the same number of turns to the first inductor.

EXAMPLE 4

In Figure 8–13, if $L_1 = 5$ H, $L_2 = 10$ H, and $L_3 = 15$ H, what is the total inductance (L_{tot}) of the circuit?

Solution:

$$L_{tot} = L_1 + L_2 + L_3$$
$$L_{tot} = 5 \text{ H} + 10 \text{ H} + 15 \text{ H}$$
$$L_{tot} = 30 \text{ H}$$

8.10 Parallel Inductors

When inductors are connected in parallel, the total inductance can be found in a similar way to finding the total resistance of a parallel circuit. The reciprocal of the total inductance is equal to the sum of the reciprocals of all the inductors in equation form:

$$L_{tot} = \cfrac{1}{\cfrac{1}{L_1} + \cfrac{1}{L_2} + \cfrac{1}{L_3} + \cdots \cfrac{1}{L_N}} \qquad (8.9)$$

where L_{tot} equals the total circuit inductance, L_1 equals the value of inductor 1, L_2 equals the value of inductor 2, L_3 equals the value of inductor 3, and L_N equals the value of the Nth inductor.

EXAMPLE 5

In Figure 8–14, if $L_1 = 5$ H, $L_2 = 10$ H, and $L_3 = 15$ H, what is the total inductance (L_{tot}) of the circuit?

Solution:

$$L_{tot} = \cfrac{1}{\cfrac{1}{L_1} + \cfrac{1}{L_2} + \cfrac{1}{L_3}}$$

$$L_{tot} = \cfrac{1}{\cfrac{1}{5} + \cfrac{1}{10} + \cfrac{1}{15}} = 2.73 \text{ H}$$

FIGURE 8–13   Inductors in series.

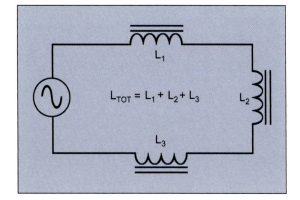

FIGURE 8–14 Inductors in parallel.

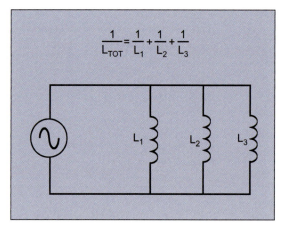

$$\frac{1}{L_{TOT}} = \frac{1}{L_1} + \frac{1}{L_2} + \frac{1}{L_3}$$

■ SUMMARY

When current flows through a conductor, a magnetic field is created around the conductor. If a conductor is cut by a magnetic field, a voltage is induced in the conductor. The polarity of the induced voltage is dictated by the polarity of the magnetic field in relation to the direction of motion. The induced voltage is always opposite to the applied voltage. This causes the inductor to oppose the change of current.

Suddenly applied DC current in an inductive circuit rises at an exponential rate. The exponential curve is divided into five time constants. The time constant is defined as the time required for the current to either increase or decrease to 63.2% of its maximum value or to decrease to 63.2% from its maximum value. Each of these calculations of the

time constant steps works to approach the full value within five time constants.

Inductance is measured in units called the henry (H) and symbolized by the letter "L."

Inductive reactance is an inductor's opposition to the flow of current in an AC circuit. Inductive reactance is measured in ohms and may be substituted into the Ohm's law formula to find other circuit parameters.

When inductors are connected in series, the total inductance is equal to the sum of all the inductors. When they are connected in parallel, the reciprocal of the total inductance is equal to the sum of the reciprocals of all the inductors.

■ REVIEW QUESTIONS

1. What is inductance? How does it affect voltage and current in a circuit?
2. What is the relationship between direction of motion, polarity of the magnetic field, and direction of induced current when a wire is passed through the field?
3. Discuss how to calculate the following:
   a. Total inductance of inductors in series
   b. Total inductance of inductors in parallel
   c. Total inductive reactance of inductors in series
   d. Total inductive reactance of inductors in parallel

4. Discuss Lenz's law.
   a. What is it, and what does it say?
   b. How does it affect the way inductors work in a circuit?
5. What physical factors affect the inductance of a coil?
6. How can you increase the inductance of a coil? Decrease it?
7. If you put a DC voltage source in series with an inductor, how much current will flow?
8. How do frequency and inductance affect inductive reactance?

■ PRACTICE PROBLEMS

1. Counter EMF always _____ the source voltage.

2. A 5-H inductor experiences a current change of 14 amperes in 7 seconds. What is the induced voltage?

3. The transient current in an inductive DC circuit is maximum after _____ time constants.

4. To calculate the total inductance for inductances in series, you _____ the individual inductances.

5. If $L_1 = 6$ H, $L_2 = 11$ H, and $L_3 = 4$ H, calculate the total inductances for the three circuits shown below.

6. If each circuit is connected to a 100-V, 60-Hz source, how much current will flow?

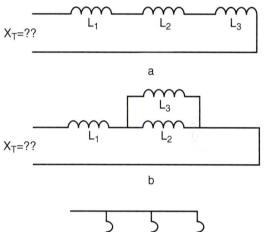

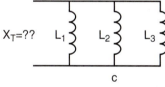

chapter 9

Inductive Reactance

■ **OUTLINE**

■ OVERVIEW

As discussed in previous chapters, an inductor has no effect on a DC circuit as long as the current remains constant. An inductor does affect any circuit where the current is changing. The inductor resists a change in current. This concept is very similar to when a magnetic field is moved around a conductor and a voltage is induced. In an AC circuit, the current is constantly changing (this also applies to fluctuating DC). This change is the expanding and collapsing of the magnetic field from zero to peak values. With these changes, the magnetic field becomes equivalent to flux in motion. The flux cutting across the conductor induces a voltage. The voltage being induced across the inductor opposes the current change. This resistance, or opposition, is called *inductive reactance* (X_L) and is measured in ohms.

This chapter reviews some of the material you learned in Chapter 8 and introduces a few new concepts. Be sure to read the material carefully and work the practice problems at the end. This material is critical to your ongoing education and training.

■ OBJECTIVES

After completing this chapter, you should be able to:

1. Determine the inductive reactance of an inductor in an AC circuit.
2. Use the inductive reactance (in ohms) in Ohm's law to determine the current flow in a purely inductive circuit.
3. Determine the value of the inductor when the voltage, current, and frequency are known.

■ GLOSSARY

Quality factor (Q) Q is the ratio of inductive reactance to resistance in an AC circuit. Q gives a measure of frequency response and the quality of behavior of a given coil as an inductor.

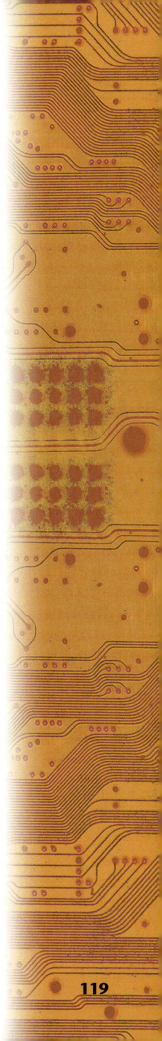

■ INDUCTIVE REACTANCE

Inductive reactance is the opposition to current flow in an AC circuit caused by the presence of inductance. In an AC circuit that has only inductance, the amount of current that flows is determined by the counterelectromotive force of the coil and is called *inductive reactance*. Notice in Figure 9–1 that the applied voltage and induced voltage are 180° out of phase. The resulting voltage is the difference between their magnitudes. Clearly, the inductor is opposing the current flow.

In Figure 9–2, the sine wave shows that at point A, the current is zero and there is no flux. From point A to point B, the positive direction of current builds to a peak value in a counterclockwise direction

FIGURE 9–1   Applied voltage versus induced voltage.

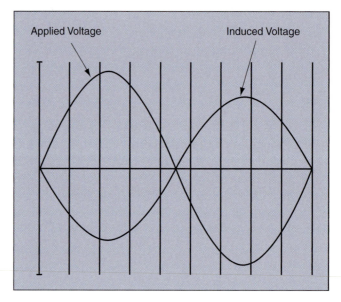

FIGURE 9–2   Induction and the counterelectromotive force.

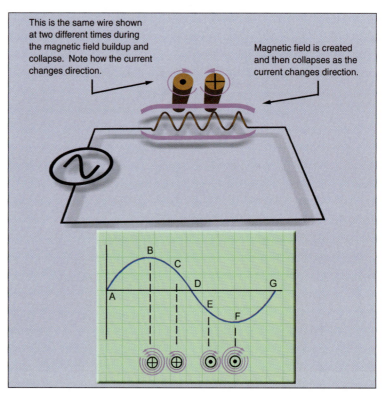

with maximum flux. At point C, the current is reducing, as is the magnetic flux. At point D, there is zero current again. The next half-cycle builds the magnetic field again, and current reverses direction. Specifically, from point D to point F, the current builds to a maximum negative peak value in a clockwise direction.

The induced voltage limits the flow of current through the circuit in a manner similar to resistance. Although the inductive reactance is not resistance, it still has the same limiting effect on current flow. The current limiting factor of the inductor is the reactance. The symbol for reactance is the letter "X." Since this reactance is caused by the inductance, X_L is used to label it. Inductive reactance is measured in ohms the way resistance is. It can be calculated when the values of inductance and frequency are known from the following formula:

$$X_L = 2\pi f L \tag{9.1}$$

Where:

X_L = inductive reactance in ohms
F = frequency in hertz
L = inductance in henrys
2π = a constant.

■ CALCULATING *L* AND *X*$_L$

The reactance of an inductive circuit depends on the inductance and the frequency. Given the frequency and inductance of a circuit, the inductive reactance can be determined.

EXAMPLE 1

Determine the amount of inductive reactance in the circuit shown in Figure 9–3.

Solution:

$$X_L = 2\pi f L$$
$$X_L = 2\pi \times 60 \times 0.7 = 263.9\ \Omega$$

Since the inductive reactance is the current-limiting property of this circuit, the next step is to substitute for the value of *R* in the Ohm's law calculation:

$$I = \frac{E}{X_L}$$

$$I = \frac{120}{263.9} = 0.455\ A$$

If the inductive reactance and the frequency are known, Equation 1 can be used to calculate the inductance of the coil:

$$L = \frac{X_L}{2\pi f}$$

FIGURE 9–3   Inductive circuit.

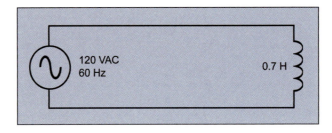

120 VAC
60 Hz

0.7 H

EXAMPLE 2

An inductor is connected to a 28-V, 400-Hz line. The circuit has a 0.5-A current flow. What is the inductance (*L*) of the inductor? What is the inductive reactance?

Solution:

$$X_L = \frac{E}{I}$$

$$X_L = \frac{28}{0.5} = 56 \ \Omega$$

Next, use the answer to the inductive reactance to determine the inductance:

$$L = \frac{X_L}{2\pi f}$$

$$L = \frac{56}{2\pi f} = 22.3 \text{ mH}$$

EXAMPLE 3

What will the current reading be if the frequency in the circuit shown in Figure 9–4 is changed to 400 Hz?

Solution:
The first step is to determine the amount of inductive reactance of the coil. Use Ohm's law:

$$X_L = \frac{E}{I}$$

$$X = \frac{440}{20} = 22 \ \Omega$$

Next, determine the inductance of the coil at 200 Hz:

$$L = \frac{X_L}{2\pi f}$$

$$L = \frac{22}{2\pi \times 200} = 17.5 \text{ mH}$$

FIGURE 9–4   Inductive circuit with a current reading.

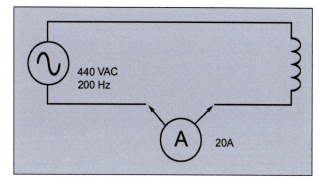

Since the inductance of the coil is determined by its physical construction, inductance will not change when connected to a different frequency. The inductance of the coil is known and has been calculated, so the 400-Hz inductive reactance can be calculated from

$$X_L = 2\pi f L$$
$$X_L = 2\pi \times 400 \times 17.5 \text{ mH} = 43.98 \ \Omega$$

The last step is to use Ohm's law to determine the current by substituting the inductive reactance value for resistance:

$$I = \frac{E}{X_L}$$

$$I = \frac{440}{43.98} = 10.00 \text{ A}$$

■ Q OF AN INDUCTOR

The calculations so far have been simplified by assuming that the inductor has no resistance. Since inductors are coils of wire, the wire has some internal resistance. Previous lessons showed that this resistance comes from various factors, including the wire size (cmils) and type of material. The ratio of the inductive reactance to the resistance of the coil is called the **quality factor** (or *Q*) of the coil. A higher *Q* means a higher-quality coil.

An inductor made of large wire will have a higher *Q* than an inductor with the same number of turns but made of a smaller wire. The formula for calculating *Q* is

$$Q = \frac{X_L}{R} \tag{9.2}$$

EXAMPLE 4

What is the *Q* of a circuit with a 0.058-H coil at 400 Hz if the coil has a resistance of 15 ohms?

Solution:
Since $X_L = 2\pi f L$,

$$X_L = 2\pi \times 400 \times 0.058 = 145.77 \ \Omega$$

$$Q = \frac{X_L}{R} = \frac{145.77 \ \Omega}{15 \ \Omega} = 9.72 \ \Omega$$

Since inductive reactance and resistance are both in ohms, the Q factor is unitless. Once the ratio of inductive reactance becomes 10 times as great as resistance, the amount of resistance is considered negligible. Therefore, a Q of 10 or greater is considered a pure inductor.

SUMMARY

Inductance is measured in a unit called the henry (H) and is symbolized by the letter "L." Inductive reactance is the opposition produced by a countervoltage that limits the flow of current similar to resistance. Inductive reactance is measured in ohms. It is proportional to the inductance of the coil and the frequency of the supply voltage. X_L symbolizes inductive reactance. In Ohm's law, X_L can be substituted for resistance in calculations.

All inductors contain some resistance. The Q of an inductor is the ratio of the inductive reactance to the resistance. When this ratio is 10 or greater, the inductor is considered to have only inductive reactance.

REVIEW QUESTIONS

1. Describe how the inductive reactance is changed by the inductance and the frequency of the circuit.

2. Compare the way that resistance (R) and inductive reactance (X_L) affect current flow in an AC circuit. In a DC circuit.

3. What will happen to the current in a circuit with only inductance if the inductive reactance is doubled? If the inductance is doubled?

4. An inductor is being chosen to limit the current flow in a certain circuit. How will the Q of the inductor affect the size chosen?

5. What is the minimum value of Q that will allow you to ignore the resistance of an inductor?

PRACTICE PROBLEMS

1. If 50 amps flow through an inductor at 100 volts, $X_L =$ _____.

2. If the circuit in question 1 is operating at 60 Hz, what is the inductance of the coil?

3. A tuning coil in a radio has an L of 300 µH. At what frequency will it have an X_L of 3,768 Ω?

4. A choke coil is used to limit the current to 50 mA at 25 V and 400 Hz. What is the inductance of the coil?

5. What is the inductance of a coil that produces 942 Ω of reactance at 60 kHz?

6. What is the current in the circuit?

7. If the inductor in question 6 has a resistance of 154 ohms, what is the Q of the inductor?

8. Will the resistance in the coil in question 7 have a significant effect on the circuit current?

9. A 10-H and a 12-H coil are used to limit current when connected in a series. What is the total inductance?

10. If the coils in question 1 were connected in parallel, what would L_T equal?

11. What would the X_L of each coil in question 1 be if they were operated at 60 Hz? What is the total reactance if the coils are in series and parallel?

12. Using the answers from questions 9 and 10, prove your answers in question 11.

chapter 10

Inductors in Series and/or Parallel

■ **OUTLINE**

INDUCTORS CONNECTED IN SERIES   **INDUCTORS CONNECTED IN PARALLEL**

126

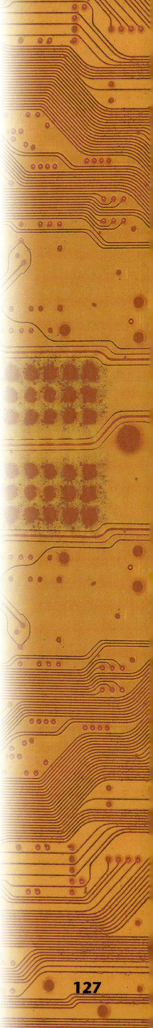

■ OVERVIEW

The ability of one coil to induce a voltage into another coil is called *mutual induction.* This happens when two coils are placed close together. A good example of this type of induction is in the windings of a transformer. Alternating current through one coil (inductor) of the transformer causes voltage to be induced into the other inductor because of the expanding and collapsing magnetic field of the first coil (see Figure 10–1).

This chapter covers the way in which inductors react in a circuit when they are connected in parallel or in series. The formulas in this chapter depend on each inductor acting independently of the others in the circuit. The inductors must be connected in the circuit so that the magnetic field from each one does not induce voltage and current into the others. The voltage that a coil induces into itself is caused by the characteristic of **self-inductance**.

■ OBJECTIVES

After completing this chapter, you should be able to:

1. Solve for total inductance in a series or parallel circuit containing more than one inductor.
2. Find total inductive reactance in a series or parallel circuit containing more than one inductor.
3. Solve for unknowns in circuits containing more than one inductor in a series or parallel when the values of other variables are specified.

■ GLOSSARY

Self-inductance   The property of an electrical component (such as a coil of wire) to induce a voltage into itself as the current through the component changes.

FIGURE 10–1   Mutual inductance.

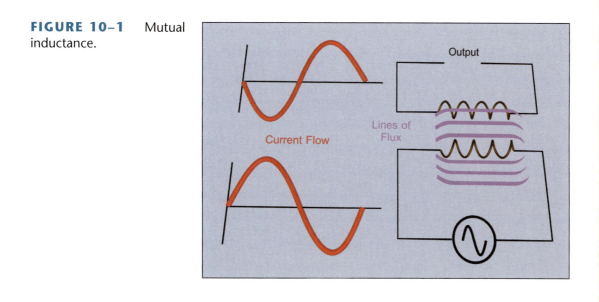

Output

Current Flow

Lines of Flux

■ INDUCTORS CONNECTED IN SERIES

When inductors are connected in series, the total inductance is the sum of the inductances of all the inductors:

$$L_T = L_1 + L_2 + L_3 + \ldots L_N \tag{10.1}$$

where N is the total number of inductors in the circuit.

It is easy to show that the total inductive reactance in a series inductive circuit is also equal to the sum of the individual reactances. Start by remembering the formula for calculating the inductive reactance if the inductance and frequency are known:

$$L = \frac{X_L}{2\pi f} \tag{10.2}$$

Substituting Equation 2 into Equation 1 yields

$$\frac{X_{L_T}}{2\pi f} = \frac{X_{L_1}}{2\pi f} + \frac{X_{L_2}}{2\pi f} + \frac{X_{L_3}}{2\pi f} + \cdots \frac{X_{L_N}}{2\pi f} \tag{10.3}$$

Multiplying Equation 3 by $2\pi f$ gives

$$X_{L_T} = X_{L_1} + X_{L_2} + X_{L_3} + \ldots X_{L_N} \tag{10.4}$$

EXAMPLE 1

The circuit in Figure 10–2 shows three inductors connected in series. L_1 has an inductance of 0.3 H, L_2 has an inductance of 0.5 H, and L_3 has an inductance 0.6 H. What is the total inductance of the circuit?

Solution:

$$L_T = L_1 + L_2 + L_3$$
$$L_T = 0.3 \text{ H} + 0.5 \text{ H} + 0.6 \text{ H} = 1.4 \text{ H}$$

FIGURE 10–2 Inductors in series.

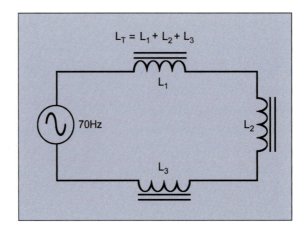

EXAMPLE 2

The inductors in Figure 10–2 (L_1, L_2, and L_3) have inductive reactance of 132 Ω, 220 Ω, and 264 Ω. What is the total inductive reactance of the circuit?

Solution:

$$X_{L_T} = X_{L_1} + X_{L_2} + X_{L_3}$$

$$X_{L_T} = 132\ \Omega + 220\ \Omega + 264\ \Omega = 616\ \Omega$$

Adding inductors in series with a circuit that already contains inductors will increase the total inductive reactance of the circuit. Also, remember from Chapter 9 that inductive reactance is measured in ohms (like resistance) and can be substituted for R in Ohm's law.

EXAMPLE 3

What is the frequency of the circuit for the inductors in Examples 1 and 2?

Solution:
The inductive reactance and the inductance are related by frequency:

$$X_L = 2\pi fL$$

Solving this equation for f and inserting the total inductance and inductive reactance for the previous examples yields

$$f = \frac{X_{L_T}}{2\pi L_T} = \frac{616\ \Omega}{2\pi \times 1.4\ \text{H}} = 70\ \text{Hz}$$

EXAMPLE 4

Assume the series circuit in Figure 10–2 has a 48-V power supply with a 400-Hz signal. What is the current?

Solution:

$$X_L = 2\pi fL$$
$$X_L = 2\pi \times 400 \times 1.4$$
$$X_L = 3{,}518.6\ \Omega$$

Substituting X_L for R in Ohm's law gives

$$I = \frac{E}{X_L}$$

$$I = \frac{48\ \text{V}}{3{,}518.6\ \Omega} = 13.6\ \text{mA}$$

EXAMPLE 5

What is the value of the current if the frequency is dropped to 30 Hz?

Solution:

$$X_L = 2\pi \times 30 \times 1.4$$

$$X_L = 263.9 \ \Omega$$

$$I = \frac{E}{X_L} = \frac{48 \text{ V}}{263.9 \ \Omega} = 18.2 \text{ mA}$$

■ INDUCTORS CONNECTED IN PARALLEL

When inductors are connected in parallel, the total inductance can be found using a formula that is similar to the one used for finding total resistance of a parallel circuit. The reciprocal of the total inductance is equal to the sum of the reciprocals of all the inductors (see Figure 10–3).

Adding additional inductors in parallel with a circuit that already contains inductors will decrease the total inductive reactance of the circuit. This is based on the principle of placing resistors in parallel. The total opposition to current flow in the circuit containing parallel inductors is always less than the opposition caused by the lowest-value coil or inductor because each time another element is added in parallel, the current has another path.

The total inductance formula for two inductors in parallel is formatted the same way as the one for two resistors in parallel:

$$L_T = \frac{L_1 \times L_2}{L_1 + L_2} \tag{10.5}$$

EXAMPLE 6

Two coils are connected in a circuit in parallel. They have an inductance of 10 H and 5 H. What is the total inductance of the two in parallel? Use the product-over-sum method.

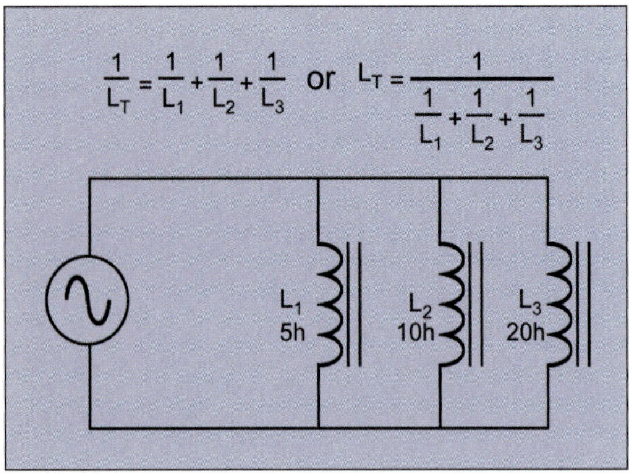

FIGURE 10–3   Inductors in parallel.

$$\frac{1}{L_T} = \frac{1}{L_1} + \frac{1}{L_2} + \frac{1}{L_3} \quad \text{or} \quad L_T = \frac{1}{\dfrac{1}{L_1} + \dfrac{1}{L_2} + \dfrac{1}{L_3}}$$

L_1 5h L_2 10h L_3 20h

Solution:

$$L_T = \frac{5 \text{ H} \times 10 \text{ H}}{5 \text{ H} + 10 \text{ H}} = 3.33 \text{ H}$$

EXAMPLE 7

What is the inductive reactance of the 10-H coil if a 400-Hz signal is applied to the circuit?

Solution:

$$X_{L_T} = 2\pi f L = 2\pi \times 400 \times 10 = 25.133 \text{ k}\Omega$$

EXAMPLE 8

What is the inductive reactance of the 5-H coil?

Solution:

$$X_{L_T} = 2\pi \times 400 \times 5 = 12.566 \text{ k}\Omega$$

EXAMPLE 9

Find the total inductive reactance of the circuit. Remember that these are in parallel. Inductive reactance for parallel circuits is found using the same formula as inductance.

Solution 1:

$$X_{L_T} = \frac{25.133 \text{ k}\Omega \times 12.566 \text{ k}\Omega}{25.133 \text{ k}\Omega + 12.566 \text{ k}\Omega} = 8.378 \text{ k}\Omega$$

Solution 2:
An alternate method would be to use the L_T found in Example 6 as follows:

$$X_{L_T} = 2\pi f L = 2\pi 400 \times 3.33 = 8.378 \text{ k}\Omega$$

EXAMPLE 10

If the power supply is set to 24 V, what is the total circuit current?

$$I_T = \frac{E}{X_{L_T}} = \frac{24}{8,378 \text{ }\Omega} = 2.86 \text{ mA}$$

EXAMPLE 11

If the frequency of the circuit in Example 10 were changed to 60 Hz, how would this affect the current and inductive reactance?

Solution:

$$X_{L_T} = 2\pi f L = 2\pi \times 60 \times 3.33 = 1{,}256.6 \ \Omega$$

$$I_T = \frac{E}{X_{L_T}} = \frac{24 \text{ V}}{1{,}256.6 \ \Omega} = 19 \text{ mA}$$

■ SUMMARY

In this chapter, you practiced solving for the total inductance in a series or parallel circuit containing more than one inductor. You also learned how to find total inductive reactance in a series or parallel circuit containing more than one inductor. Each inductor has to be physically independent from other inductors in the circuit for these formulas to work; that is, their magnetic fields must not interfere with each other. Like resistances in series and parallel, other circuit unknowns are calculated to make a complete circuit analysis.

When inductors are connected in series, the total inductance is equal to the sum of all the inductors. When the inductors are connected in parallel, the reciprocal of the total inductance is equal to the sum of the reciprocals of all the inductors.

■ REVIEW QUESTIONS

1. Consider the following figure. If all the circuit elements were inductors instead of resistors, you might wish to calculate the power dissipation in watts for each element. What sort of answer would you get if you multiplied the current times the voltage for each element in the following figure?

2. How do you calculate the total inductance for inductors in parallel? In series?

3. How do you calculate the total inductive reactance for inductors in parallel? In series?

4. How is the inductive reactance of an inductive circuit changed if the frequency goes up? Goes down?

5. If you know the inductive reactance, can you calculate the inductance? What do you need to know first?

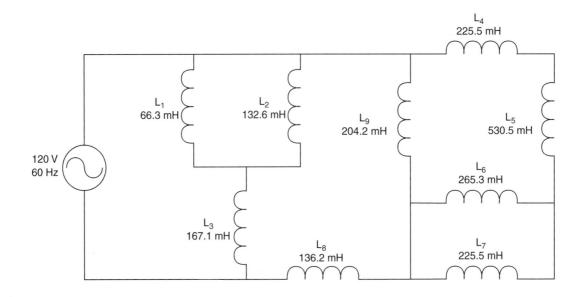

■ PRACTICE PROBLEM

1. In the circuit shown on the previous page, calculate the following:

   a. Reactance of each inductor (you should round the answers to the nearest ohm)

   b. Total reactance of the circuit

   c. Total inductance of the circuit

   d. Current through each inductor

   e. Voltage drop across each inductor

PART

4

CAPACITANCE IN AC CIRCUITS

chapter 11

Capacitance and Its Effect on Circuits

OUTLINE

■ OVERVIEW

So far in your training, you have learned a great deal about resistance and resistive circuits, and you have learned the fundamentals of inductance in circuits. In this chapter and the next two, you will learn about the third member of the passive element team: the capacitor.

Inductance, as you learned, opposes a change in the current flow. It does this by creating a counterelectromotive force when the current tries to change. Capacitance is the complement of inductance. Capacitance opposes a change in voltage by creating a current flow that opposes the change in voltage.

Capacitors are used in electrical circuits of all kinds. They are used to block DC and pass AC, to correct power factor problems in systems, to set timing constants for various types of oscillators, and for a variety of other applications.

Generally, capacitor operation is not as intuitive as inductor operation. The reasons for this are beyond the scope of this training; however, there is some good news. The formulas, concepts, and circuit analysis used when capacitors are involved are very similar to those involving inductors. The differences are less important than the similarities, as you will soon see.

■ OBJECTIVES

After completing this chapter, you should be able to:

1. Define capacitance, its unit of measure, and the factors that affect it.
2. Explain the operation of AC and DC circuits that contain capacitance.
3. Calculate the capacitance based on the basis of the electrical or physical quantities given.

GLOSSARY

Capacitance   1. The ratio of charge to potential on an electrically charged, isolated conductor. 2. Symbol C. The ratio of the electric charge transferred from one to the other of a pair of conductors to the resulting potential difference between them. 3. The property of a circuit element that permits it to store charge. The part of the circuit exhibiting capacitance.[1]

Capacitor   An electric circuit element used to store charge temporarily, consisting in general of two metallic plates separated and insulated from each other by a dielectric. Also called **condenser.**[2]

Dielectric   A nonconductor of electricity, especially a substance with electrical conductivity less than a millionth (10^{-6}) of a siemens.[3]

Farad   The unit of capacitance in the meter-kilogram-second system equal to the capacitance of a capacitor having an equal and opposite charge of 1 coulomb on each plate and a potential difference of 1 volt between the plates.[4]

Microfarad 1×10^{-6} farads. Abbreviated as μF.

Micromicrofarad   See Picofarad. Abbreviated as $\mu\mu$F.

Nanofarad 1×10^{-9} farads. Abbreviated as nF.

Picofarad 1×10^{-12} farads. Abbreviated as pF. Also called a micromicrofarad (abbreviated as $\mu\mu$F).

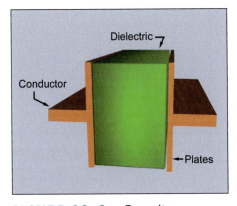

FIGURE 11–1   Capacitor construction.

■ CAPACITOR CONSTRUCTION AND RATINGS

11.1 Physical Characteristics

Capacitors perform a variety of jobs, such as storing an electrical charge to produce a large current pulse, timing circuits, or power factor correction. Look at Figure 11–1. A capacitor consists of two conductors, usually plates, separated by some type of insulating material, called the **dielectric**. The insulating material can also be air. Capacitors are devices that oppose a change in voltage.

Three factors affect the **capacitance** of the capacitor: the area of the plates, the distance between the plates, and the type of dielectric used. The formula for calculating the capacitance of a capacitor is

$$C = \frac{x\varepsilon A}{d} \times 10^{-13} \qquad (11.1)$$

Where:

C = capacitance in **farads**
ε = dielectric constant of the material between the plates (for air, $\varepsilon = 1$)
A = cross-sectional area of the plates (in cm^2 or in.^{2})
d = distance between the plates (in cm or in.)
x = 0.0885 if A and d are in centimeters
x = 0.225 if A and d are in inches.

The unit of capacitance is covered in more detail later in this chapter.

To charge a capacitor, you connect it to a DC power source, and electrons will be removed from the side connected to the positive pole and then deposited on the side of the capacitor connected to the negative pole (see Figure 11–2). This flow will continue until the voltage across the capacitor equals the voltage across the source. When these two voltages become equal, the flow will stop, and the capacitor is charged. A good rule to know about capacitors and current is that current flow can occur or change only during the time when the capacitor is charging or discharging.

In theory, the capacitor should remain charged forever. In actual practice, it will not. No dielectric is perfect. The electrons will eventually move from the negative plate to the positive plate through the dielectric, causing the capacitor to discharge. This current flow is called *leakage current* and is inversely proportional to the resistance of the dielectric and directly proportional to the voltage across the plate. If the dielectric of a capacitor becomes weak, it will permit an unacceptable amount of current flow. Such a capacitor is called a *leaky capacitor*.

11.2 Electrostatic Charge

Capacitance is the ability of a circuit component to store an electrical charge. In previous chapters, you learned that a coil has a constantly moving electromagnetic field. A capacitor, on the other hand, has a

FIGURE 11–2 Battery connected to a capacitor; electrons build up on the negative side.

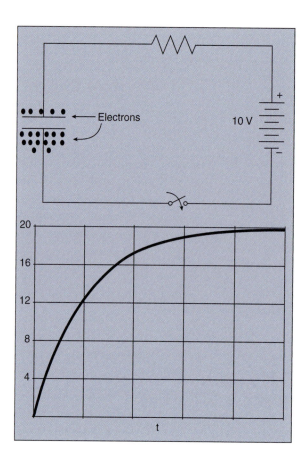

changing electrostatic field. The term *electrostatic* refers to the fact that it has an electrical charge that is stationary. This is the charge that is produced when the electrons are removed from one plate and deposited on the other. This storage of a potential difference, or voltage, causes tension and a pull on the insulating material. Think of this as a slingshot being drawn back. When the two sides of the capacitor are shorted, the electrons are "fired," and the tension between the two sides of the capacitor is released, returning the dielectric material to neutral (see Figure 11–3).

11.3 Capacitor Rating

The unit used for capacitance is the **farad** and is symbolized by an (F). The magnitude of capacitance is determined by the formula

$$C = \frac{Q}{V}$$

FIGURE 11–3 Effect of electrostatic charge.

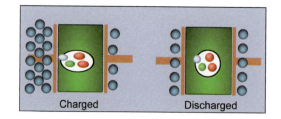

Charged   Discharged

Where:

> Q is the charge in coulombs
>
> V is the potential difference (voltage) between the conductors (plates)
>
> C = the capacitance in farads.

A capacitor has the capacitance of 1 farad when a change of 1 volt across its plates results in the movement of 1 coulomb of charge. A farad, however, is a very large amount of capacitance, so smaller units or decimal fractions of the farad are used. These units are the **microfarad** (μF), **nanofarad** (nF), and **picofarad** (pF):

$$\text{Microfarad} = \mu\text{F} = \frac{1}{1,000,000} = (1 \times 10^{-6}) \text{ of a farad, or } 0.000001 \text{ F}$$

$$\text{Nanofarad} = \text{nF} = \frac{1}{1,000,000,000} = (1 \times 10^{-9}) \text{ of a farad, or } 0.000000001 \text{ F}$$

$$\text{Picofarad} = \text{pF} = \frac{1}{1,000,000,000,000} = (1 \times 10^{-12}) \text{ of a farad, or } 0.000000000001 \text{ F}$$

The picofarad is also called the **micromicrofarad** and is symbolized by $\mu\mu$F.

11.4 Voltage and Current Relationship in a Capacitor

You previously learned that the relationship between the voltage drop on an inductor is given by the formula

$$E_L = -L\frac{\Delta I_L}{\Delta t} \tag{11.2}$$

Where:

> E_L = the voltage across the inductor
>
> I_L = the current through the inductor
>
> L = equals the inductance in henries.

You also learned that the term $\Delta I/\Delta t$ is the change in current with respect to the change in time. A capacitor has a very similar formula that relates its voltage and current:

$$I_C = C\frac{\Delta E_C}{\Delta t} \tag{11.3}$$

Where:

> E_C = the voltage across the capacitor
>
> I_C = the current through the capacitor
>
> C = equals the capacitance in farads.

$\Delta E_C/\Delta t$ = the change in capacitor voltage with respect to the change in time. Both of these formulas will be very important later when you learn about the phase shifts associated with capacitors and inductors.

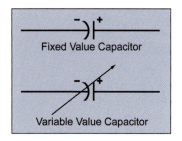

FIGURE 11–4   Capacitor symbols.

11.5 Schematic Symbols

The schematic symbol is a straight line opposed by an arc (see Figure 11–4).

11.6 Voltage Ratings

The voltage rating of the capacitor is actually the rating of the dielectric. The voltage rating is extremely important for the life of the capacitor and should never be exceeded. The voltage rating indicates the maximum amount of voltage that the dielectric is intended to withstand without breaking down. If the voltage becomes too great, the dielectric will break down, allowing current to flow between the plates. In this condition, the capacitor is referred to as *shorted.* Since the AC voltage rating is an RSM value, the actual peak value of the applied voltage on the capacitor will be considerably higher. You can calculate this value by multiplying the AC voltage by an RMS conversion value of 1.414.

For example, a 28-VAC value applied to a capacitor would require the capacitor to have at least a peak voltage rating of $28 \times 1.414 = 39.59$ V.

■ CAPACITORS AND CURRENT

Remember that a capacitor is made from two metal plates separated by an insulating material called a dielectric. The insulator prevents the flow of electrons through the capacitor. When connected to alternating voltage supply, current appears to flow through the capacitor. The reason is that in an AC circuit, the polarity is continually changing, and the current is changing direction.

To demonstrate, look at Figure 11–5. There are two rooms connected by a blower. Periodically, the blower switches directions, and the airflow goes in the opposite direction. The blower represents the AC power source, and the rooms represent the plates of the capacitor. The airflow will move back and forth but never make a complete circuit through both rooms in one direction.

Current cannot flow through a capacitor but does flow into and out of it each cycle. This constitutes current flow through a load connected in the circuit without having current flow through the capacitor.

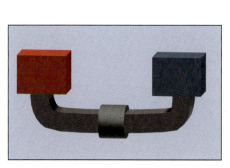

FIGURE 11–5   Two rooms with a reversing blower.

11.7 Charge and Discharge Rates

Capacitors charge and discharge at an exponential rate, the curve of which is divided into five time constants. During each time constant the voltage will change by 63.2% of the amount left to reach the fully charged state. During the first time constant, the voltage across the capacitor will reach 63.2% of the supply voltage measured across it. During the next time constant, the capacitor will reach a higher voltage equal to 63.2% of the voltage left between the first time constant voltage and the supply voltage, which is then added to the voltage reached in the first time constant.

For example, assume that the capacitor in Figure 11–3 requires 10 seconds to reach the supply voltage of 20 volts. The capacitor can

Table 11–1   Charging Values for the Capacitor in Figure 11–2

Time Constant	Calculation	Final Voltage
1	63.2% of 20 V = 12.64	12.64 V
2	(20 − 12.64) × 63.2% = 4.65	12.64 + 4.65 = 17.29 V
3	(20 − 17.29) × 63.2% = 1.71	17.29 + 1.71 = 19.00 V
4	(20 − 19.00) × 63.2% = .632	19.00 + .632 = 19.632 V
5	(20 − 19.632) × 63.2% = .232	19.632 + .232 = 19.864 V

charge only to the system supply voltage. Table 11–1 shows the voltages across the capacitor during the 10 seconds.

As you can see, by the end of the fifth time constant, the voltage across the capacitor has reached 99.3% of the total voltage. If this continues past the fifth time constant, the voltage will continue to get closer to the final value; however, it will never reach the maximum. Notice that the voltage charge on the capacitor behaves in exactly the same way as the current flow through an inductor. Figure 11–6 shows the entire curve.

The capacitor discharges in the same manner. At the end of the first time constant, the voltage will decrease by 63.2% of its charged value. The voltage will continue to drop until it reaches approximately zero, when five time constants have elapsed (see Figure 11–7).

Knowing that five time constants are required to reach the maximum travel is only half the picture. Of equal importance is determining how long a time constant is in seconds. The length of one time constant is dependent on the size of the capacitor and the resistance in series with the capacitor and is given by the formula

$$\tau = R \times C \tag{11.4}$$

Where:

τ = one time constant in seconds

R = circuit resistance in ohms

C = circuit capacitance in farads.

The Greek letter tau (τ) is used to designate the time required for one time constant. The capital letter "T" also is used.

EXAMPLE 1

How long will it take a circuit with a 10-kΩ resistor and a 20-μF capacitor to complete one time constant? How long will it take for the capacitor to fully charge?

Solution:

$\tau = R \times C$

$\tau = 10\ \text{k}\Omega \times 20\ \mu\text{F} = 0.2\ \text{seconds}$

To completely charge

$T_{\text{tot}} = 5 \times \tau = 1\ \text{second}$

FIGURE 11–6   Time constants for charging a capacitor.

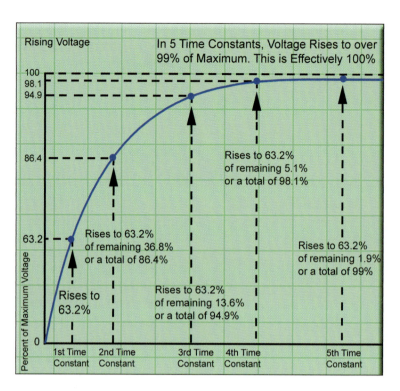

FIGURE 11–7   Time constants for a discharging capacitor.

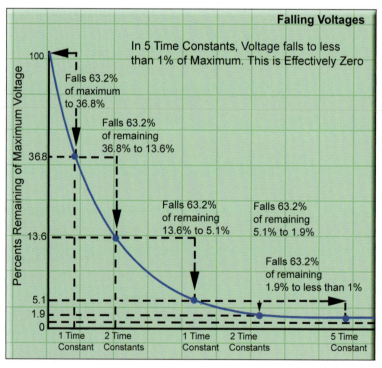

■ DIELECTRIC CHARACTERISTICS

11.8 Dielectric Stress

When a capacitor is charged, a potential exists between the plates of the capacitor. The plate with the lack of electrons has a positive charge, and the plate with the excess has a negative charge. The molecules in the dielectric are like small electrical dipoles; that is, they are more

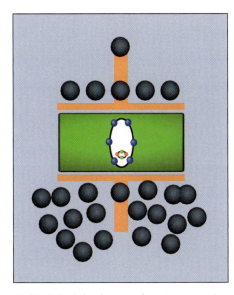

FIGURE 11–8 Dielectric stress in a charged capacitor.

positive on one side than they are on the other. The negatively charged plate attracts the positive side of the molecules, and the positively charged plate attracts the negative side of the molecules. This attraction causes the molecules to twist and stretch. They do not actually move to any great extent, but, like a strong spring or rubber band, they will store energy. This action stores energy in the dielectric of a charged capacitor. It has little effect on the capacitor during AC operations; however, it does have some ramifications during DC testing of insulation (see Figure 11–8).

11.9 Dielectric Constants

The type of dielectric used is an important factor in determining the amount of capacitance a capacitor will have. Materials are assigned a number called the *dielectric constant,* usually abbreviated with an uppercase "K" or the Greek letter eta (ε). Air is assigned the number 1 and is used as the reference point for comparison. Table 11–2 shows the dielectric constants for various materials. A change in the dielectric of the capacitor will change its rating. For instance, a capacitor with an air dielectric (1) and a capacitor rating of 5 μF would change its rating if the dielectric changed to Teflon (2):

$$5 \text{ μF} \times 2 \text{ (dielectric rating)} = 10 \text{ μF}$$

Table 11–2 Dielectric Constants of Different Materials

Material	Dielectric Constant
Air	1
Bakelite	4–10
Castor oil	4.3–4.7
Cellulose acetate	7
Ceramic	1,200
Dry paper	3.5
Hard rubber	2.8
Insulating oils	2.2–4.6
Lucite	2.4–3.0
Mica	6.4–7.0
Mycaflex	8.0
Paraffin	1.9–2.2
Porcelain	5.5
Pure water	81
Pyrex glass	4.1–4.9
Rubber components	3.0–7.0
Teflon	2
Titanium dioxide compounds	90–170

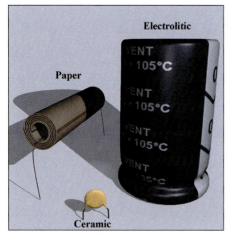

FIGURE 11–9 Types of commercial capacitors.

Figure 11–9 shows some different types of commercial capacitors that you may encounter during your career as an electrical worker.

■ CAPACITIVE CURRENT, VOLTAGE, AND PHASE RELATIONSHIPS

In a pure capacitive circuit, the current leads the applied voltage by 90°. To understand this, you need to recognize two significant points:

1. Applying an AC sinusoidal voltage to the capacitor will result in an AC sinusoidal current.
2. The current through the capacitor is determined by the change in the voltage across the capacitor: $i = C(\Delta V/\Delta t)$.

Look at Figure 11–10. Table 11–3 shows the $\Delta V/\Delta t$ for every 10° of the first 90° of the sine wave. Notice that the current will be at its greatest when the voltage is changing the fastest and vice versa. Also notice that, as stated previously, if the voltage is a sine wave, the current will be a sine wave as well. Since the change in the voltage is greatest when the sine wave is close to zero, you would expect the current to be greatest when the voltage is close to zero. This is shown by looking at Figure 11–11. The figure shows the current through a capacitor and the

FIGURE 11–10 A sine wave: (a) 0° to 360°; (b) 0° to 40°; (c) 40° to 90°.

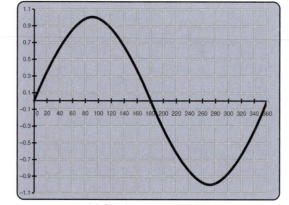

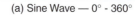

(a) Sine Wave — 0° - 360°

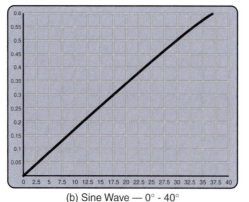

(b) Sine Wave — 0° - 40°

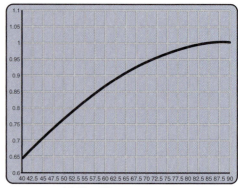

(c) Sine Wave — 40° - 90°

Table 11–3 Change in Voltage for Change in Time of the First 90°

Interval	1	2	3	4	5	6	7	8	9
Δt	0–10	10–20	20–30	30–40	40–50	50–60	60–70	70–80	80–90
$\dfrac{\Delta V}{\Delta t}$	.1736	.1684	.1580	.1428	.1233	.0999	.0737	.0451	.0152

FIGURE 11–11 Voltage across versus current through a capacitor.

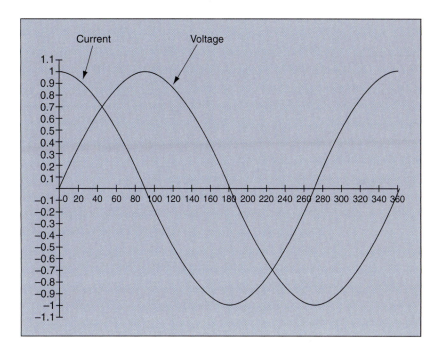

voltage across the capacitor. Notice three things about these waveforms:

1. Both are sine waves.
2. The current is at a maximum when the voltage is changing the fastest. Remember that a sine wave changes the fastest when it is close to zero.
3. The voltage is lagging the current by 90°.

This analysis shows that the voltage across a capacitor will lag the current through the capacitor by 90°.

■ FREQUENCY EFFECTS

Frequency has an effect on capacitive reactance because the capacitor charges and discharges faster at a higher frequency. Recall that current is the rate of electron flow. If the frequency is increased, the charge will flow faster because the capacitor is being charged and discharged faster (see Figure 11–12). Since the capacitor is being charged and discharged

FIGURE 11–12 Capacitor charge at 30 Hz and 60 Hz.

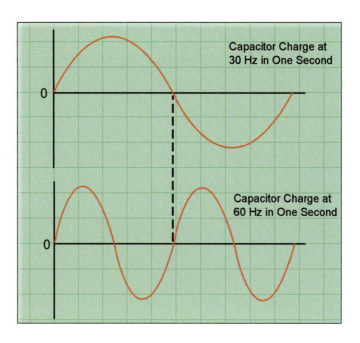

faster, the resistance to current flow is decreased, so current flow will increase according to the following:

$$I = Q/t$$

Where:

I = current

Q = charge in coulombs

t = time in seconds.

If a capacitor is connected to a 30-Hz line, 1 coulomb of charge flows per second, and the frequency is doubled to 60 Hz, 1 coulomb will flow in ½ second because the capacitor is charging and discharging twice as fast. Thus, the opposition to current flow is decreased. Compare the sine waves in Figure 11–12.

■ SUMMARY

In this chapter, you learned that capacitors are devices that oppose a change of voltage. Three factors determine the capacitance of a capacitor: the surface areas of the plates, the distance between the plates, and the type of dielectric. Energy is stored in an electrostatic field in the capacitor. Current flows in a capacitor as the plates charge and discharge. The time required to charge or discharge a capacitor is measured in time constants. It takes five time constants to charge or discharge a capacitor.

The basic unit of capacitance is the farad. The farad is a large unit, so decimal portions called microfarads, nanofarads, and picofarads (or micromicrofarads) are used.

When a capacitor is connected to an alternating current, current appears to flow because of the continuous increase and decrease in voltage and the changing polarity of the AC source. Current does not flow through the capacitor. Voltage across a capacitor lags the current through the circuit by 90°.

REVIEW QUESTIONS

1. In your own words, what is capacitance? Relate it to charge and voltage.
2. What is a nanofarad? A picofarad? Why are these terms used?
3. Explain how a capacitor stores a charge.
4. Describe the way in which voltage will build up on a capacitor.
   a. How fast does it build up?
   b. What is its maximum value?
5. What is a time constant?
6. How do the following affect a capacitor?
   a. Plate area
   b. Plate separation
   c. Dielectric constant of the insulation
7. What is the phase angle relationship between the current through and the voltage on a capacitor?
8. In a capacitive current, how does frequency of the applied signal affect the following?
   a. Charging rate
   b. Opposition to current flow

PRACTICE PROBLEMS

1. Given $C = K (A/D) \times 8.85 \times 10^{-12}$, find the capacitance ($C$) of a two-plate capacitor if the area of one plate is .0025 cm^2 and the separation is 0.02 cm ($K = \varepsilon = 7$).
2. What is the capacitance of a capacitor that stores 4 C of charge at 2 V?
3. What is the charge taken on by a 10-F capacitor at 3 V?
4. What is the voltage across a .001-F capacitor that stores 2 C?
5. A DC source of voltage has a 10-M Ω resistance in series with a 1-μF capacitor. What is the time constant?
6. Find the resistance necessary in a series resistive inductive (RC) circuit if the circuit has a capacitance of 10μF and a time constant of 1 second is desired.
7. How long will it take for the capacitor in question 6 to fully charge?
8. If the source voltage is 100 V, what is the voltage across the capacitor after one time constant? After five time constants?
9. List three factors that determine the capacitance of a capacitor.
10. Where does the capacitor store its energy?
11. The capacitance of a capacitor with air as a dielectric constant is .248 μF. Find the new capacitance if rubber is used instead of air with a constant of 3, the plate area is increased by ¼, and the plate separation is reduced by ¾ of its original value.

chapter 12

Capacitive Reactance

■ OUTLINE

CAPACITIVE REACTANCE

12.1 Capacitive Reactance Formula
12.2 Review of Units of Measurement

CALCULATING CAPACITANCE

■ OVERVIEW

Like resistors and inductors, capacitors also oppose the flow of current. Like inductors, the capacitor's opposition to current flow varies with frequency. This opposition that a capacitor exhibits is called capacitive reactance. The unit of capacitive reactance is the ohm, and it will replace resistance in Ohm's law for purely capacitive circuits.

In an inductive circuit, current lags voltage (think "ELI": current [I] comes after voltage [E] in an inductive [L] circuit). You will remember that inductive reactance is directly proportional to frequency; conversely, capacitive reactance is inversely proportional to frequency. In a capacitive circuit, voltage lags current (think "ICE": voltage [E] comes after current [I] in a capacitive [C] circuit).

This chapter will extend your knowledge of the behavior of capacitors. In particular, you will learn how to calculate the capacitive reactance knowing the capacitance of the capacitor and the frequency of the applied voltage.

■ OBJECTIVES

After completing this chapter, you should be able to:

1. Explain the concept of capacitive reactance and the factors that affect it.
2. Calculate capacitive reactance, frequency, and capacitance when any two of these three factors are known.

■ GLOSSARY

Capacitive reactance   The opposition to current flow by a capacitor. The symbol for capacitive reactance is X_C. The formula to calculate it is $X_C = \dfrac{1}{2\pi fC}$.

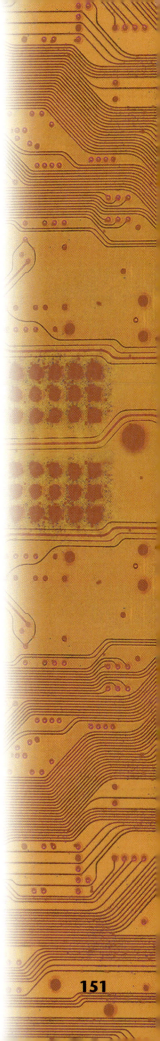

■ CAPACITIVE REACTANCE

When a capacitor is charged, a potential difference (voltage) is developed across the plates as the electrostatic charge builds up. The potential difference is the voltage provided by the electrostatic charge. This voltage opposes the applied voltage and limits the flow of current in the circuit (see Figure 12–1). You may remember that this is similar to the opposing voltage, or countervoltage, in an inductive circuit caused by an inductor.

12.1 Capacitive Reactance Formula

Look at the circuit in Figure 12–1. When the switch is closed, electrons will flow from the source to the bottom plate of the capacitor. At the same time, electrons are flowing from the top plate of the capacitor back to the source. As this happens, voltage builds up across the plates and opposes the voltage source.

This countervoltage is also called *reactance,* and since it is caused by a capacitor, it is called **capacitive reactance**. The symbol for capacitive reactance is X_C, and it is measured in ohms. The formula for calculating capacitive reactance is

$$X_C = \frac{1}{2\pi f C} \tag{12.1}$$

Where:

X_C = capacitive reactance
π = 3.14159
f = frequency in hertz
C = capacitance in farads.

12.2 Review of Units of Measurement

You have learned that because of the size of some electrical parameters, there is a need for decimal equivalents or prefix multipliers. Such equivalents are required when the actual unit is very small or very

FIGURE 12–1   Capacitor opposition to electron flow.

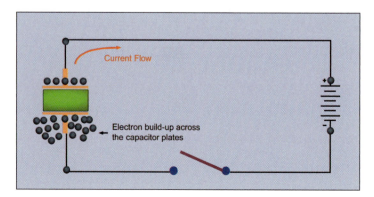

Current Flow

Electron build-up across
the capacitor plates

Table 12–1   Common Multipliers and Their Symbols

Unit	Number	Symbol	Example	Power of 10
Pico	$\dfrac{1}{1,000,000,000,000}$	p	$1\ pF = \dfrac{1}{1,000,000,000,000}\ F$	$\times 10^{-12}$
Micromicro[1]	$\dfrac{1}{1,000,000,000,000}$	$\mu\mu$	$1\ \mu\mu F = \dfrac{1}{1,000,000,000,000}\ F$	$\times 10^{-12}$
Nano	$\dfrac{1}{1,000,000,000}$	n	$1\ nV = \dfrac{1}{1,000,000,000}\ V$	$\times 10^{-9}$
Micro	$\dfrac{1}{1,000,000}$	μ	$1\ \mu F = \dfrac{1}{1,000,000}\ F$	$\times 10^{-6}$
Milli	$\dfrac{1}{1,000}$	m	$1\ mV = \dfrac{1}{1,000}\ V$	$\times 10^{-3}$
Kilo	$1,000$	k	$1\ kV = 1,000\ V$	$\times 10^{3}$
Mega	$1,000,000$	M	$1\ M\Omega = 1,000,000\ \Omega$	$\times 10^{6}$

[1]Remember that micromicro and pico are equal to each other. The former is an older term that has fallen out of use.

large. Table 12–1 lists several of the most common multipliers. Here are some examples of how these multipliers are used:

- Power plants produce megawatts.
- The appliances in our homes use kilowatts.
- The electronic circuits in your computer use milliwatts.
- The frequency of the human voice varies from hertz to kilohertz.

Chapter 11 discussed microfarads, picofarads (also called micromicrofarads), and nanofarads as a measure of capacitance.

EXAMPLE 1

A capacitor circuit has a 35-μF capacitor connected to a 48-V, 60-Hz line. How much current will flow through this line?

Solution:

First find the capacitive reactance. The C in the formula is in farads, so the conversion from μF to farads is times 10^{-6}. Thus, the line has 35×10^{-6} farads:

$$X_C = \frac{1}{2\pi f C}$$

$$X_C = \frac{1}{2 \times 3.14159 \times 60 \times (35 \times 10^{-6})}$$

$$X_C = \frac{1}{0.013195}$$

$$X_C = 75.79\ \Omega$$

Now use Ohm's law to calculate the current:

$$I = \frac{E}{X_C}$$

$$I = \frac{48}{75.79}$$

$$I = 0.63 \text{ A}$$

■ CALCULATING CAPACITANCE

Now that you have calculated capacitive reactance, you can determine the capacitance of any capacitor. Note that capacitance is also dependent on frequency and X_C. The formula is

$$C = \frac{1}{2\pi f X_C}$$

EXAMPLE 2

A capacitor is connected to a 220-V, 60-Hz line. The ammeter on the line indicates 3.2 A. What is the capacitance value of the capacitor?

Solution:

The first step is to calculate capacitive reactance using Ohm's law.

Since

$$X_C = \frac{E}{I}$$

then

$$X_C = \frac{220 \text{ V}}{3.2 \text{ A}}$$

$$X_C = 68.75 \ \Omega$$

$$C = \frac{1}{2\pi f X_C}$$

$$C = \frac{1}{2 \times 3.14159 \times 60 \times 68.75}$$

$$C = \frac{1}{25{,}918}$$

$$C = 3.86 \times 10^{-5}, \text{ or } 38.6 \ \mu\text{F}$$

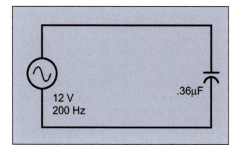

FIGURE 12–2 Capacitor circuit for example 3.

EXAMPLE 3

Look at Figure 12–2. What is the reactance of a 0.36-μF capacitor connected to a source that delivers 12 V at 200 Hz?

Solution:

$$X_C = \frac{1}{2\pi fC}$$

$$X_C = \frac{1}{2 \times 3.14159 \times 200 \times .00000036}$$

$$X_C = \frac{1}{.00045238}$$

$$X_C = 2210.5 \ \Omega$$

What is the current in this circuit?

$$I = \frac{E}{X_C}$$

$$I = \frac{12}{2210.5}$$

$$I = 0.00543 \ A, \text{ or } 5.4 \ mA$$

■ SUMMARY

In a capacitive circuit, the current appears to flow through the capacitor because of the continuous increase and decrease of voltage and the continuous change in polarity in an AC circuit. The current flows back and forth in the circuit but does not flow through the capacitor. Electrons build up on one side of the capacitor and then on the other side as current alternates. Figure 12–3 illustrates current flow for half this cycle.

FIGURE 12–3 Capacitor electron flow.

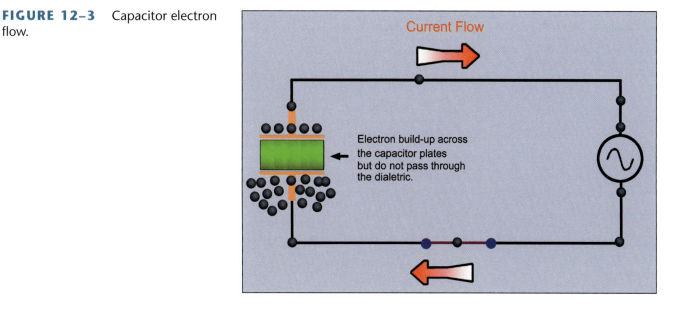

The current flow in the capacitive circuit is limited by capacitive reactance. Capacitive reactance is inversely proportional to the capacitance of the capacitor and the frequency of the line voltage source.

Capacitive reactance is similar to inductive reactance, as it is measured in ohms and can be substituted for resistance in the Ohm's law formula. In a pure capacitive circuit, the current leads the applied voltage by 90°.

If the frequency of a circuit is increased, the capacitive reactance is decreased. If the capacitance is decreased, the capacitive reactance is increased.

■ REVIEW QUESTIONS

1. Discuss the concept of capacitive reactance.
   a. What is it?
   b. How does it change if the capacitance increases?
   c. How does it change if the frequency decreases?
   d. How is it used in Ohm's law?

■ PRACTICE PROBLEMS

1. A 20-µF capacitor produces a voltage drop of 5 V at 1 KHz. What is the current?

2. What is the capacitance of a capacitor that has a reactance of 800 Ω at 10 KHz?

3. A 20-pF capacitor draws 10 mA at 95 V. What is the frequency?

4. A decrease in frequency will show a ____ in reactance.

5. If frequency is increased, the current will ____.

6. If frequency is increased, the voltage drop will ____.

7. An 80-µF capacitor is connected to a 240-V, 60-Hz supply. What is the current?

chapter 13

Capacitors in Series and/or Parallel

In previous chapters, you learned how to solve series, parallel, and combination circuits using resistors and inductors. The calculations for solving capacitor circuits are similar to those of resistance and inductance circuits. The primary differences are that capacitors in series are calculated like resistors and inductors in parallel and that capacitors in parallel are calculated like resistors and inductors in series.

■ **OBJECTIVES**

After completing this chapter, you should be able to:

1. Explain how capacitance acts on a circuit when its components are connected in series, in parallel, or in combinations.
2. Mathematically solve for the total capacitance and/or capacitive reactance in a circuit.

◼ CAPACITORS IN SERIES

Connecting capacitors in series effectively increases the distance between the plates, thereby reducing the total capacitance of the circuit. The total capacitance for series capacitors can be calculated in a similar fashion to parallel resistance.

When inductors are connected in parallel, the total inductance can be found in a similar way to finding the total resistance of a parallel resistive circuit. The reciprocal of the total inductance is equal to the sum of the reciprocals of all the inductors:

$$L_T = \frac{1}{\dfrac{1}{L_1} + \dfrac{1}{L_2} + \dfrac{1}{L_3}} \quad \text{or} \quad \frac{1}{L_T} = \frac{1}{L_1} + \frac{1}{L_2} + \frac{1}{L_3}$$

When resistors are connected in parallel, there are several ways to calculate total resistance. When all the resistors are of equal value, the total resistance is equal to the resistance of one of the resistors or branches divided by the number of resistors or branches:

$$R_T = \frac{R}{N}$$

If two resistors of unequal resistance are in parallel, the formula for the total is

$$R_T = \frac{R_1 \times R_2}{R_1 + R_2} \tag{13.1}$$

If more than two unequal resistors are in parallel, the formula is

$$R_T = \frac{1}{\dfrac{1}{R_1} + \dfrac{1}{R_2} + \dfrac{1}{R_3}}$$

By placing capacitors in series, the total dielectric is made thicker, as though you took the dielectric of each of the capacitors and placed them in series between one set of plates. Consequently, the effective distance between the plates increases, thus reducing the total value of capacitance in the circuit. In other words, the distances between the plates are added (see Figure 13–1). In this figure, you see three capacitors connected in series. Look at the formula given, where C equals the amount of capacitance (measured in picofarads), K is a dielectric constant based on the dielectric material used to make up the capacitor, A represents the surface area of the plates (measured in square inches), and d represents the distance between plates (measured in inches). When this formula is applied to capacitors in series, the effective distance d between the plates is increased. You can see that as the distance increases, the capacitance becomes smaller.

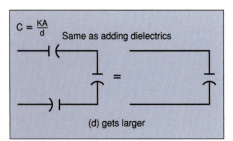

FIGURE 13–1 Capacitors connected in series.

13.1 Series Capacitor Formulas

The formulas for capacitance in a series circuit are as follows:

$$C_T = \frac{1}{\dfrac{1}{C_1} + \dfrac{1}{C_2} + \dfrac{1}{C_3}}$$

$$C_T = \frac{C_1 \times C_2}{C_1 + C_2}$$

$$\frac{1}{C_T} = \frac{1}{C_1} + \frac{1}{C_2} + \frac{1}{C_3}$$

$$C_T = \frac{C}{N} \qquad (13.2)$$

Where:

C = capacitance of one capacitor

N = number of equal capacitors connected in series.

The last formula (13.2) cannot be used unless all the individual series capacitors are equal.

EXAMPLE 1

What is the total capacitance of three capacitors connected in series with the values of $C_1 = 20\ \mu F$, $C_2 = 25\ \mu F$, and $C_3 = 45\ \mu F$?

Solution:

$$C_T = \frac{1}{C_1} + \frac{1}{C_2} + \frac{1}{C_3}$$

$$C_T = \frac{1}{\dfrac{1}{20} + \dfrac{1}{25} + \dfrac{1}{45}}$$

$$C_T = \frac{1}{0.05 + 0.04 + 0.022}$$

$$C_T = \frac{1}{0.112}$$

$$C_T = 8.91\ \mu F$$

13.2 Calculating Voltage Drops for Series Capacitors

Since the current is the same at any point in a series circuit, the voltage drop across each capacitor can be computed using the capacitive reactance of each capacitor and the current flow of the circuit.

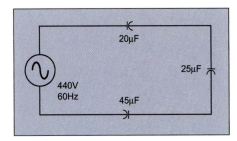

FIGURE 13–2 Series-connected capacitor circuit for examples.

EXAMPLE 2

Calculate the voltage drops for each of the capacitors in Figure 13–2.

Solution:

Step 1

Find the capacitive reactance of each capacitor:

$$X_{C_1} = \frac{1}{2 \times \pi \times 60 \times (20 \times 10^{-6})}$$

$$X_{C_1} = \frac{1}{.007536}$$

$$X_{C_1} = 132.7 \ \Omega$$

$$X_{C_2} = \frac{1}{2 \times \pi \times 60 \times (25 \times 10^{-6})}$$

$$X_{C_2} = \frac{1}{.00942}$$

$$X_{C_2} = 106.1 \ \Omega$$

$$X_{C_3} = \frac{1}{2 \times \pi \times 60 \times (45 \times 10^{-6})}$$

$$X_{C_3} = \frac{1}{.0016956}$$

$$X_{C_3} = 58.9 \ \Omega$$

Note that since $\frac{1}{2\pi}$ is a constant, you can substitute the equivalent value and simplify the formula to the following:

$$X_C = \frac{0.159}{fC}$$

Step 2

Now find the total capacitive reactance:

$$X_{C_T} = X_{C_1} + X_{C_2} + X_{C_3}$$

$$X_{C_T} = 132.7 + 106.1 + 58.9$$

$$X_{C_T} = 297.7 \ \Omega$$

Step 3

Now find the current in the circuit:

$$I_T = \frac{E_{C_T}}{X_{C_T}}$$

$$I_T = \frac{440}{297.7}$$

$$I_T = 1.48 \ A$$

Step 4

Now find the voltage drop across each capacitor:

$E_C = I_C \times X_C$ (for any capacitor)

$E_{C_1} = 1.48 \times 132.7$

$E_{C_1} = 196.4$ V

$E_{C_2} = 1.48 \times 106.1$

$E_{C_2} = 157.0$ V

$E_{C_3} = 1.48 \times 58.9$

$E_{C_3} = 87.2$ V

Remember that to check this, you can add the voltage drops together. The sum of the individual voltage drops should be equal to the total voltage from the supply:

$E_T = E_{C_1} + E_{C_2} + E_{C_3}$

$E_T = 196.4$ V $+ 157.0$ V $+ 87.2$ V

$E_T = 440.6$ V

The .6 volts of the final answer, 440.6 V, is caused by math round-off errors throughout the calculations.

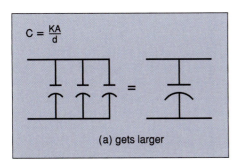

FIGURE 13–3   Capacitors in parallel.

■ CAPACITORS IN PARALLEL

When the capacitors are connected in parallel, the areas of the individual plates are added together, increasing the capacitance. Connecting capacitors in parallel has the same effect as increasing the plate area of one capacitor. Look at Figure 13–3. The formula used to determine the capacitance when the dielectric constant K, the area of the plates A (in square inches), and the distance between the plates d (in inches) also applies here. You can see that increasing the surface area of the plates increases the capacitance value C. Capacitors in parallel are calculated similar to inductors and resistors in series.

You learned that when inductors are connected in series, the total induction of the circuit equals the sum of the inductances of each of the inductors:

$$L_T = L_1 + L_2 + L_3 \quad \text{(series)}$$

The total resistance in a series circuit is found by adding the values of the resistors in the circuit:

$$R_T = R_1 + R_2 + R_3 \quad \text{(series)}$$

The total capacitance of a parallel capacitance circuit is the following formula:

$$C_T = C_1 + C_2 + C_3 \quad \text{(parallel)}$$

EXAMPLE 3

Given the circuit in Figure 13–4, with capacitors of 20 µF, 30 µF, and 60 µF, calculate the total capacitance.

Solution:

$$C_T = C_1 + C_2 + C_3$$
$$C_T = 20 + 30 + 60$$
$$C_T = 110 \ \mu F$$

EXAMPLE 4

What is the voltage drop across each capacitor in Figure 13–4?

Solution:
Remember that the voltage drop across each component connected in parallel is equal to the voltage drop across the supply. The voltage drop is therefore 110 volts:

$$E_T = E_{C_1} = E_{C_2} = E_{C_3} = 110 \ V$$

EXAMPLE 5

If the supply voltage in Figure 13–4 has a frequency of 50 Hz, what will be the total X_C for the circuit? What will be the total current?

Solution:

$$\text{Total } X_{C_{tot}} = \frac{1}{2\pi f C_{tot}}$$

$$\text{Total } X_{C_{tot}} = \frac{1}{2 \times 3.14159 \times 50 \times (110 \times 10^{-6})}$$

$$\text{Total } X_{C_{tot}} = \frac{1}{0.03456}$$

$$\text{Total } X_{C_{tot}} = 28.94 \ \Omega$$

FIGURE 13–4   Parallel capacitor circuit for examples.

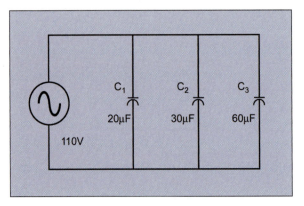

Solution:

$$I_T = \frac{V_T}{X_{C_{tot}}}$$

$$I_T = \frac{120}{28.94}$$

$$I_T = 4.15 \text{ A}$$

EXAMPLE 6

Calculate the total capacitive reactance of Figure 13–4 by using the individual capacitor reactances.

Solution:
Reactances in parallel behave the same as resistances in parallel. Therefore, the total reactance is

$$\frac{1}{X_{C_T}} = \frac{1}{X_{C_1}} + \frac{1}{X_{C_2}} + \frac{1}{X_{C_3}}$$

$$X_{C_1} = \frac{1}{2\pi fC} = \frac{1}{.0063}$$

$$X_{C_2} = \frac{1}{2\pi fC} = \frac{1}{.0094}$$

$$X_{C_3} = \frac{1}{2\pi fC} = \frac{1}{.0188}$$

$$\frac{1}{X_{C_T}} = .0063 + .0094 + .0188 = .035$$

$$X_{C_T} = 28.94 \ \Omega$$

■ COMBINATION CAPACITANCE CIRCUITS

EXAMPLE 7

Given the combination circuit in Figure 13–5 and using the formulas you have learned, solve for total circuit capacitance (C), capacitive reactance (X_C), and current (I).

Solution 1 (Total Capacitance):
First, calculate the total capacitance of the two capacitors in parallel (C_2 and C_3) because they will end up being in series with the C_1 and C_4:

$$C_{parallel} = C_2 + C_3$$

$$C_{parallel} = 4 \ \mu F + 8 \ \mu F$$

$$C_{parallel} = 12 \ \mu F$$

Now use the reciprocal formula for capacitors in series and treat the combined value of C_2 and C_3 as if it were in series with the other two capacitors:

$$C_T = \frac{1}{\dfrac{1}{C_1} + \dfrac{1}{C_4} + \dfrac{1}{C_{2,3}}}$$

$$C_T = \frac{1}{\dfrac{1}{6} + \dfrac{1}{5} + \dfrac{1}{12}} = 2.22 \ \mu F$$

Solution 2 (Capacitive Reactance):
What is the total capacitive reactance for this circuit?

$$X_{C_{tot}} = \frac{1}{2\pi f C} = \frac{1}{2\pi \times 60 \times 2.22 \ \mu F} = 1,194.8 \ \Omega$$

Solution 3 (Circuit Current):

$$I = \frac{E}{X_{C_{tot}}} = \frac{48}{1,194.8} = 40.2 \ mA$$

FIGURE 13–5   Capacitor combination circuit.

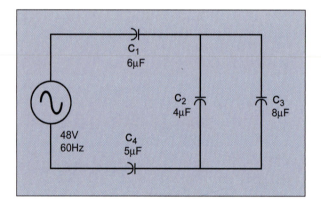

■ SUMMARY

In this chapter, the calculations for series, parallel, and combination capacitance circuits were discussed and put into practice. The total capacitance of series capacitors is calculated the same way as the total resistance of parallel resistors, and the total capacitance of parallel capacitors is calculated the same way as the total resistance of series resistors. The total reactance of any combination of capacitors is calculated in the same way as resistances and inductive reactances.

When you connect two identical capacitors in series, you effectively double the thickness of the di-electric in the circuit, which yields a total capacitance equal to half the value of each capacitor. The current remains constant in series-connected capacitive circuits.

When capacitors are connected in parallel, it has the same effect as increasing the plate area of one capacitor. When capacitors are connected in parallel, the voltage drop is the same as the voltage drop of the supply, just as with any other components connected in parallel.

REVIEW QUESTIONS

1. What is the effect of connecting capacitors in series? What is the effective dielectric thickness? How does this affect the total capacitance?
2. What is the effect of connecting capacitors in parallel? What is the effective cross-sectional area of the plates? How does this affect the total capacitance?
3. Briefly describe how to calculate the total capacitance of the following:
   a. Capacitors in series
   b. Capacitors in parallel
4. Briefly describe how to calculate the total capacitive reactance of the following:
   a. Capacitors in series
   b. Capacitors in parallel

PRACTICE PROBLEMS

1. Find the total capacitance of three capacitors connected in series with values of 3 μF, 5 μF, and 10 μF.
2. What is the total capacitance of three 300-μF capacitors connected in series?
3. What value of capacitor must be connected in series with a 10-μF capacitor to get a total capacitance of 8 μF?
4. What is the total capacitance of two capacitors connected in parallel having values of 310 μF and 50 μF?
5. What is the X_{C_T} of a circuit having two capacitors connected in parallel with values of 300 μF and 600 μF operating at 500 Hz?
6. When two capacitors are in a series, the total X_C is _____ than the smaller capacitor.
7. When two capacitors are connected in parallel, the total capacitance is _____ than the smaller capacitor.
8. What value of capacitor must be added in parallel with a .23-μF capacitor to get a total of 5,000 Ω of X_C total in a 60-Hz, 1,000-V circuit?
9. Repeat practice problem 8 but find the value of the capacitor when connected in series with a .83-μF capacitor. What current flows in the circuit?

PART

5

COMBINING RESISTANCE, INDUCTANCE, AND CAPACITANCE

chapter 14

Characteristics of AC Circuits

■ OUTLINE

■ OVERVIEW

Previous chapters introduced you to all the fundamentals needed to understand and analyze AC circuits. In this chapter, you will review some of the material you have previously studied. It prepares you for the following chapters, which will teach in-depth methods and techniques.

■ OBJECTIVES

After completing this chapter, you should be able to:

1. Define the terms and concepts associated with AC circuits containing resistors, capacitors, and inductors.
2. Explain the effects of resistance, capacitance, and inductance on power consumption of a circuit.

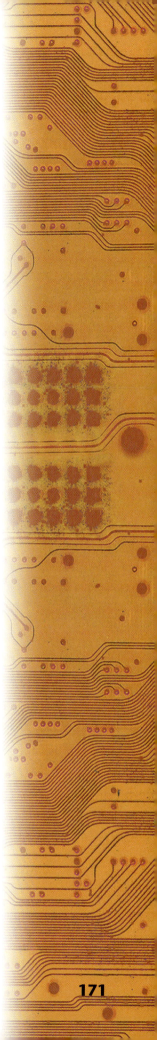

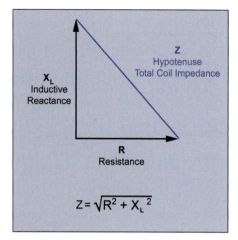

FIGURE 14–1 Coil impedance (a combination of wire resistance and inductive reactance).

■ REVIEW

When DC or AC voltage is applied across a resistor, a current flows. The relationship between voltage, resistance, and current is explained using Ohm's law (E = IR). However, circuits with resistors, inductors, and capacitors do not react the same as circuits that contain only resistors. Circuits may have resistors, capacitors, inductors, or any combination of these.

In a purely inductive circuit, when direct current flows through a coil, the only force that opposes the flow of current is the resistance of the wire from which the coil is made; however, if the same circuit is connected to an AC source, both the resistance of the wire and the reactance of the coil will oppose the current flow. Reactance is the induction of voltage that limits current flow in a coil. This reactance always opposes the original change in current. That is why the induced voltage is known as a counterelectromotive force.

■ IMPEDANCE

The total current-limiting effect of an inductor is a combination of the inductive reactance and the resistance. The combined effect of the coil resistance and the coil reactance is the total current-limiting effect, or the impedance. Impedance is symbolized by the letter "Z." Since the unit of measure for both the resistance and the reactance is the ohm, the unit of measure for impedance is also the ohm. Although both resistance and reactance oppose the flow of current, these values cannot simply be added algebraically because they are 90° out of phase with each other (see Figure 14–1).

Since resistance and reactance form the legs of a right triangle, the total opposition is equal to the value of the hypotenuse. The relationship between the resistance, reactance, and impedance, as represented by the triangle in Figure 14–1, provides the basis by which impedance is calculated. To compute the value of impedance for the coil, the inductive reactance and resistance must be added vectorially. The Pythagorean theorem is used to compute the magnitude of impedance and is represented by the formula in Figure 14–1:

$$C^2 = A^2 + B^2$$

$$Z^2 = (R)^2 + (X_L)^2$$

$$Z^2 = (5)^2 + (10)^2 \quad \text{(assumes R = 5 and } X_L = 10)$$

$$Z^2 = 25 + 100$$

$$Z = \sqrt{125}$$

$$Z = 11.18 \ \Omega$$

Ohm's law says that it takes 1 volt to push 1 ampere through 1 ohm. The formulas are

$$E = I \times R$$

$$I = \frac{E}{R}$$

$$R = \frac{E}{I}$$

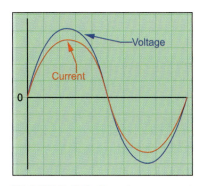

FIGURE 14–2 Current and voltage in phase (a pure resistive circuit).

Since the impedance is the total current-limiting factor of the circuit, it can be used to replace R in an Ohm's law formula. The total current is found by using the formula

$$I_T = \frac{E_T}{Z}$$

EXAMPLE 1

Given a circuit with an impedance of 7 ohms and a voltage of 110 ohms, calculate the total current.

Solution:

$$I_T = \frac{E_T}{Z}$$

$$I_T = \frac{110}{7}$$

$$I_T = 15.7 \text{ A}$$

■ PHASE ANGLES

14.1 Resistive and Inductive Phase Shifts

When an AC voltage is applied to a resistor, the current flow will be a "copy" of the voltage. This means that the current will rise and fall at the same rate and time as the voltage and will reverse in polarity at the same time and in the same direction that the voltage reverses polarity. As you can see from Figure 14–2, the voltage and current in the resistive circuit hit their zero points and peak points at exactly the same time. When the voltage is positive, the current is also positive.

The relationship between the phase angles is a function of time. If they start at the same time on the x-axis, rise and fall at the same time, and return to the x-axis in the same direction at the same time, there is no time difference between the two waveforms. In this condition, the current is said to be in phase with the voltage.

Actually, there are no purely resistive, inductive, or capacitive circuits. Although some circuits may be predominantly resistive, inductive, or capacitive, in reality all circuits have at least a small amount of all three components. This means that the current and voltage are unlikely to be exactly in phase.

In a circuit that has measurable resistance and inductance, the circuit current and voltage will always be slightly out of phase. The size of the phase angle and the number of degrees of offset or time lag will be determined by the relative size of the inductive reactance and the resistance. Larger inductive reactance (compared to the resistance) results in a larger phase angle. In an inductive circuit, the current will always lag the voltage. If it is pure inductance, the phase angle will be 90°. As resistance is added, the phase angle will decrease toward zero. An inductive phase angle is shown graphically in Figure 14–3.

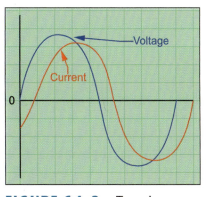

FIGURE 14–3 Two sine waves out of phase.

An easy way to remember the voltage–current relationship is found in the following mnemonic:

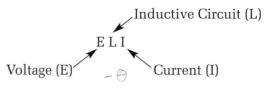

In this mnemonic, the order of the letters "E" and "I" show the phase relationship between the current and the voltage. A similar mnemonic is covered for a capacitive circuit later in this chapter.

14.2 Theta

The amount of time by which the voltage and current are out of phase with each other is called *angular displacement* and is represented by the Greek letter theta (θ). If the current and voltage are out of phase by $45°$ $\theta = -45°$. Note the minus sign indicating that the current is lagging the voltage.

The phase displacement represented by theta exists only in an AC circuit. There is no phase difference in DC circuits, where there is only one type of load: resistive. Even motors that are made up of coils of wire appear to be resistive because of their conversion of electrical energy into mechanical energy.

14.3 Capacitive Circuit Phase Shifts

In a pure inductive circuit, the current lags the voltage by 90°. The opposite voltage–current relationship appears in a purely capacitive circuit. In the capacitive circuit, the current will lead the applied voltage by 90°. This means that the voltage will be crossing the x-axis after the current when viewing the voltage and current waveforms plotted on the same graph.

An easy way to remember the voltage–current relationship is found in the following mnemonic:

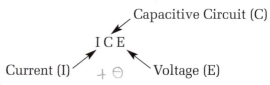

In a capacitive circuit, as indicated by the center letter "C," the current, I, leads the voltage, E. Again, the order of the letters in the mnemonic determines which leads and which lags.

The voltage and current have the same polarity for half the time and opposite polarities the other half. Look at Figure 14–4. During the period that the polarities are the same (blocks B and D), energy is being stored in the capacitor in the form of an electrostatic charge. When the voltage and current have opposite polarities (blocks A and C), the capacitor is discharging. Power is stored when the capacitor is charged, and an equal amount of power is returned to the circuit when the capacitor discharges. There is no true power produced in a pure capacitance circuit.

FIGURE 14–4   Power stored and discharged in a capacitive circuit circuit.

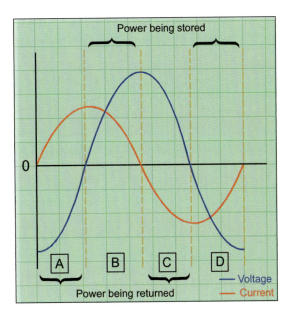

14.4 Power and Phase Effects

In reviewing power in a resistive circuit, all the power delivered by the source to the load is consumed or dissipated by the load. In an inductive circuit, only a portion of the power supplied by the source is actually dissipated by the load resistance. The balance that is not consumed by the circuit load resistance, the portion delivered to the inductor, is delivered back to the source as the field created within the inductor collapses.

The same is true in a capacitive circuit. In a circuit that contains resistance and capacitance, only the power associated with the resistive portion of the circuit is actually dissipated. The portion associated with the capacitor will actually be returned to the circuit as the capacitor discharges.

In a circuit with resistance, capacitance, and inductance, the portion of power that is actually consumed by the resistive component is called *true power*. That portion of power that is stored in the magnetic field of the inductor or electrostatic field of the capacitor and then injected back into the circuit is called *reactive power*. The vector sum of these powers is called *apparent power*.

Apparent power is calculated by multiplying the total circuit voltage by the total circuit current. This power is the portion that, to the unknowing eye, would *apparently* be used by the circuit. It would be easy to make the assumption that if a circuit is measured and found to have certain amounts of voltage and current associated with it, then the circuit must be utilizing the measured amount to do work.

When current and voltage in a circuit reach maximum positive or negative values at the same time, true power is at its maximum possible value. This could happen only if they were in phase (see Figure 14–5).

True power can be produced only when current and voltage are in phase. Apparent power, the result of multiplying the circuit voltage by the current, is measured in volt-amperes (VA) and will always be equal to or greater than the true power. When apparent and true power are equal, the voltage and current are in phase with each other.

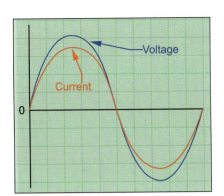

FIGURE 14–5   Voltage and current in phase.

$P_{APP} \geq P_{true}$

$V_A \geq P_{true} = E_R \times I_R$

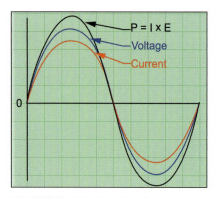

FIGURE 14–6 In-phase voltage and current.

According to Ohm's law, in a pure resistive circuit, the power can be found by multiplying the voltage times the current (see Figure 14–6):

$$P = I \times E$$

However, in a pure inductive circuit, no true power (watts) is produced. Recall that both voltage and current must be either positive or negative to achieve a positive value for power. Since the voltage and current are 90° out of phase with each other, power will be stored in the magnetic field for 50% of the time (when voltage and current have the same polarity) and returned to the circuit 50% of the time (when voltage and current have opposite polarities). Therefore, no power is actually dissipated from the circuit (see Figure 14–7). As you can see in the figure, the power is being stored during the first quarter of each current cycle (both positive and negative) and is being returned during the second quarter of each current cycle.

■ CALCULATING DIFFERENT TYPES OF POWER

14.5 Power Factor

The power factor of the circuit is a ratio of the true power to the apparent power. The value ranges from 0 to 1 and is usually presented as a percentage (0% to 100%). Power factor can be computed in any series circuit by dividing the circuit's total resistor value by the total opposition to the flow of current: the impedance. The following formulas give variations of that concept:

$$PF = \frac{E_R}{E_T} \text{ (voltage drop across the circuit resistance by the total circuit voltage)}$$

$$PF = \frac{R}{Z} \text{ (resistance divided by impedance)}$$

FIGURE 14–7 Out-of-phase power (VAR).

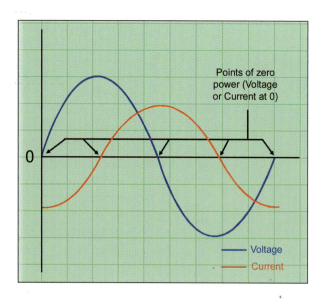

$$PF = \frac{P}{VA} \text{ (watts divided by volt-amps)}$$

PF = cos θ (cosine of the angle between the current and the voltage)

The decimal fraction is converted to a percentage by multiplying by 100. The power factor can be leading or lagging, depending on whether the circuit is capacitive (leading) or inductive (lagging). Remember that in a series circuit, the current is the same in all parts of the circuit, so the power factor cannot be calculated using current.

The power factor is important in industrial applications. Power is sold by electrical utility companies based on true power or watts consumed. The electrical utility must, however, supply the apparent power. If an industrial plant has a power factor of 65% and is consuming 5 MW of true power, the electrical utility must actually supply 7.7 MVA (megavolt-amps). This is calculated by dividing the true power by the power factor:

$$\frac{5 \text{ MW}}{.65} = 7.7 \text{ MVA}$$

Although the additional reactive power is not being "used," the current that supplies it has to travel back and forth from the utility to the industrial plant. This additional current causes additional losses that the utility must make up.

If the power factor were corrected to 95%, the power company would have to supply only 5.26 MVA to furnish the same amount of true power to the plant:

$$\frac{5 \text{ MW}}{.95} = 5.26 \text{ MVA}$$

This is why power companies charge a surcharge for plants with low power factors.

EXAMPLE 2

A steel mill has a requirement for 15 MVA and uses 11.7 MW. What is the power factor of the mill?

Solution:

$$\frac{11.7 \text{ MW}}{15 \text{ MVA}} = .78 = \text{PF } 78\%$$

Power factor is discussed in detail in Chapter 27.

14.6 Reactive Power

Reactive power in a circuit is referred to as volt-amps-reactive (VARs). VARs should not be confused with watts or true power. VARs is the product of the volts and amps that are 90° out of phase with each other. The voltage drop across an inductor or capacitor and the current through them are examples. Remember that true power can be produced only when the current and voltage are in phase with each other. VARs are often called *wattless power*.

During times when both current and voltage are either positive or negative, power is stored in the form of a magnetic (L) or an electrostatic (C) field. During times that voltages have opposite signs, the power is returned to the circuit (see Figure 14–7):

$$\text{VARs} = I_\text{L}^2 \times X_\text{L} \quad \text{or} \quad I_\text{C}^2 \times X_\text{C}$$

14.7 Apparent Power

Volt-amperes (VA) is the apparent power of the circuit. It can be calculated in a similar fashion as watts or VARs, except that the total values of current and voltage are used. It is called *apparent* power because it is the value that would be found if a voltmeter and ammeter were used to measure the circuit voltage and current and the measured values were multiplied together (see Figure 14–8):

$$\text{VA} = E_\text{T} \times I_\text{T}$$

14.8 True Power

Since true power or watts can be produced only by the in-phase components of current and power, only resistive parts of the circuit can produce watts.

True power, measured in watts, is equal to apparent power when the total current and total voltage are in phase, the phase angle is 0, and the power factor is 100%. Power factor, however, can range from 0% to 100%. The impact of the phase angle on this I^2R power rating can be calculated as

$$E_\text{T} \times I_\text{T} \times \cos \theta$$

where θ can be calculated by

$$\tan \theta = \frac{X_\text{L}}{R} = \frac{E_\text{L}}{E_\text{R}}$$

FIGURE 14–8   Apparent power as the product of measured values.

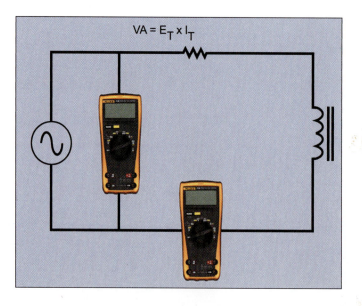

VA = E_T x I_T

Multiplying $E_T \times I_T$ by the cosine of the phase angle provides the resistive component for true power. To prove this, look at Figure 14–9.

Power factor:

$$\tan \theta = \frac{X_R}{R}$$

$$\tan \theta = \frac{173}{100}$$

$$\tan \theta = 1.73$$

$$\theta = 59.98°$$

$$PF = \cos \theta$$

$$PF = \cos 59.98 = .5 = 50\%$$

$$\text{True power} = E \times I \times \cos \theta$$

$$400 \text{ V} \times 2 \text{ A} \times .5 = 400 \text{ W}$$

True power could also be calculated by I^2R for $R_1 = (2 \text{ A})^2 \times 100 = 400$ W. Apparent power is simply calculated by $V \times I = 400$ V $\times$ 2 A = 800 VA.

Knowing both true and apparent power, the power factor can be calculated as

$$PF = \text{true power (W)/apparent power (VA)} = \frac{400 \text{ W}}{800 \text{ VA}} = .5 = 50\%$$

The power factor could have also been calculated by

$$PF = \frac{R}{Z}$$

$$PF = \frac{100}{\sqrt{(100^2 + 173^2)}}$$

$$PF = \frac{100}{199.8} = .5 = 50\%$$

Recall that $Z = \sqrt{R^2 + X_L^2}$.

FIGURE 14–9 Calculating true power.

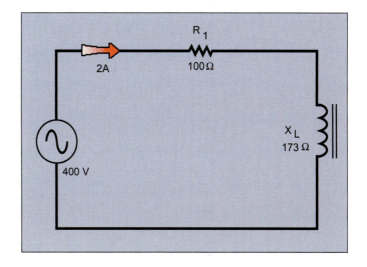

■ ELECTRICAL EQUIPMENT RATED IN KVA (APPARENT POWER) VERSUS WATTS (TRUE POWER)

Many loads are mainly resistive by nature and therefore are rated in watts (true power). Simply stated, these loads have a very small reactive component, so an apparent power rating (VA) would not be appropriate. If you recall, apparent power (VA) is both the true power (resistive component) and the reactive power (inductor and capacitor components) of a circuit. Examples of resistive loads are incandescent lightbulbs, water heaters, unit heaters, hair blowers, and cooking ranges.

Transformers are rated in kilovolt-amperes (KVA) because apparent power represents the total current (and voltage) that the transformer can supply. The total full-load amps that a 25-KVA transformer can deliver to a single-phase load at 240 volts is 104 amps. This is calculated by the power equation $I = P/E$. It is up to the electrician to ensure that the connected load to the transformer (apparent or true) does not exceed the transformer's full current rating.

Generator sets are mechanical devices made up of an engine turning the shaft of a generator that supplies power to a load. The generator is usually rated in kilowatts (KW) with a power factor of 80%. All generators are considered to be rated at 80% power factor unless the nameplate states differently. The KW rating of a generator requires the load to be examined in respect to KW (true power). This approach allows the electrician to accurately size the generator.

For example, a generator is to serve a building that is made up of several different types of loads: lighting panels of resistive lighting (higher power factor), lighting panels for fluorescent lighting (lower power factor), power panels feeding HVAC with motors (poor power factor), receptacle panels feeding general office equipment (lower power factor), and so on. As you can see, the building's power factor is difficult to calculate at any one time.

If a generator were rated at 100 KVA with a total building load of 100 KW, a building power factor of 80% would actually create a 125-KVA load. This generator would be undersized for the load. By rating the generator at 100 KW with an assumed 80% power factor, the chance of incorrectly sizing a generator set is less. Unlike a transformer, which provides power by means of mutual inductance, a generator is directly coupled to an engine. A load larger than the generator's maximum KW rating will physically load the engine so that it cannot deliver the horsepower required by the generator.

The KVA rating of a generator can be calculated by dividing the KW by the power factor (KVA = KW/.8). But, as mentioned previously, the tendency to use the calculated KVA of a generator will often result in an undersized generator for the load served.

■ SUMMARY

This chapter reviewed some of the fundamental concepts for resistive, inductive, and capacitive circuits.

When a steady direct current flows through a coil, the only opposition is the resistance from the wire in the coil. When alternating current flows through a coil, it is opposed by two factors: resistance and reactance.

Apparent power (measured in VA) is simply the multiplication of the circuit's total current by the voltage. It includes both true power (resistive power measured in watts) and reactive power (reactance power measured in VARs). In theory, purely inductive circuits contain no true power (or watts) because of the absence of the resistive component.

True power (measured in watts) is the actual power used by the load or as the resistive part of the circuit that performs the work. True power can be produced only when current and voltage are in phase. In theory, we discuss purely resistive circuits, but in real situations they do not exist because the circuit conductors themselves produce some amount of both inductance and capacitance.

Reactive power is measured in VARs, which is an abbreviation for "volts-amps-reactive." It is the power that flows back into the source from the inductors (and capacitors). It is this opposing power that affects the power factor of the circuit. Even though this opposing reactive power is not desired, without inductors (magnetism) we would not have motors and transformers.

The power factor is a ratio of the true power (watts) to apparent power (VA). An increase in reactive power (VARs) causes the power factor to decrease. The decreased power factor means increased wasted power used to perform the work. The utility company will penalize the customer for a poor power factor because of the wasted power. In later chapters, several additional formulas will be introduced to calculate apparent power, true power, reactive power, and power factor.

The current and voltage in a pure resistive circuit are in phase with each other. The current and voltage in a pure capacitive circuit are 90° out of phase with each other. The current and voltage in a pure inductive circuit are 90° out of phase with each other. Induced voltage is always opposite in polarity to the applied voltage. Inductors oppose a change of current.

The total opposition to flow of AC through a circuit is called impedance, which is measured in units called ohms and represented by the letter "Z." Impedance can be substituted for resistance in Ohm's law.

■ REVIEW QUESTIONS

1. What are apparent power, true power, and reactive power, and what is their relationship to each other?
2. What kind of power is involved for the following phase angles between current and voltage?
   a. 0
   b. 90
   c. −90
   d. 180
3. Explain the meaning of the mnemonic "ELI the ICE man" that was discussed in chapter 12.
4. List and discuss all the various formulas that are used to calculate power factor.
5. $Z = \sqrt{R^2 + X^2}$. Show this formula graphically and explain what it means.

■ PRACTICE PROBLEMS

1. In an impedance triangle, what does the horizontal vector represent?
2. What is the impedance of a circuit that has a 20-Ω resistor in series with a 20-Ω X_L?
3. If 20 volts are supplied to the circuit in question 2, what current will flow?

The remaining questions refer to the following figure.

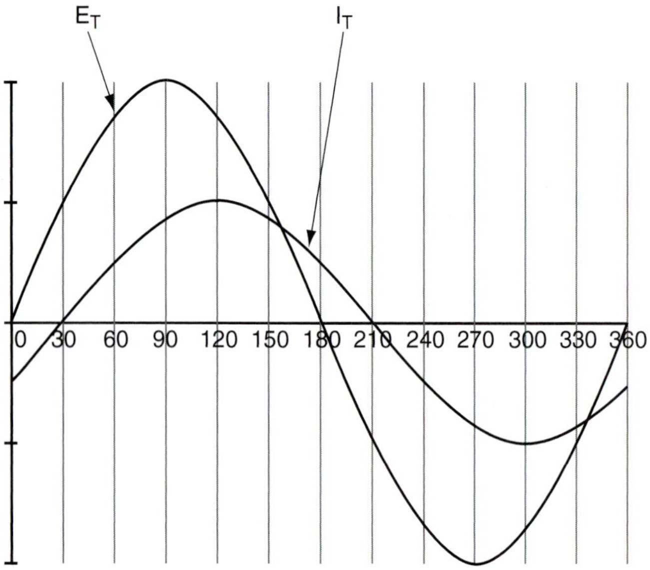

4. What is the phase angle between the total voltage and the total current in the figure?
5. Which wave is leading?
6. Is this an inductive or a capacitive circuit (Use "ELI the ICE Man" from chapter 12?)
7. Construct an impedance vector diagram.
8. What is the power factor of this circuit?
9. If the total voltage is 100 volts and the total current is .2 amps, what apparent power is being delivered to the circuit?
10. How much true power is being used by this circuit?
11. What size inductive reactance would you need to have a PF of 50% if the resistor is 23.1 Ω? What size coil is needed at 60 Hz?

chapter 15

Parameters of Series RL Circuits

■ OUTLINE

■ OVERVIEW

In a series RL circuit, current in the inductor and the resistor are the same. Since E_R is in phase with the current I_R, and E_L is 90° out of phase with the current I_L, E_L and E_R are 90° out of phase with each other. The relationship among all the vector values (R and X_L, E_R, and I_R, E_T, and Z) in a series RL circuit is shown in Figure 15–1.

This chapter reviews previously studied material. Resistance, capacitance, and inductance have all been discussed before, and most AC circuits contain two or more of these characteristics in series, parallel, or combination circuits. This chapter focuses on circuits with resistance and inductance in series. In previous lessons, you learned that the voltage (E_L) leads the current (I_L) by 90° in an inductive circuit. Conversely, in a purely resistive circuit, the voltage (E_R) and current (I_R) are in phase.

■ OBJECTIVES

After completing this chapter, you should be able to:

1. Explain the relationship between circuit parameters in a series RL circuit.
2. Solve mathematically for unknown properties in a series RL circuit.

FIGURE 15–1 Series Resistor–Inductor (RL) Relationships.

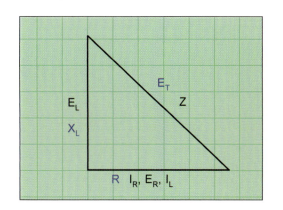

■ REVIEW

15.1 Resistance

The total resistance of a series circuit is calculated by adding the values together. The common term to remember is "resistance adds":

$$R_T = R_1 + R_2 + R_3$$

Also, remember that the general formula for calculating the total resistance of resistors in parallel is called the *reciprocal formula*:

$$\frac{1}{R_T} = \frac{1}{R_1} + \frac{1}{R_2} + \frac{1}{R_3}$$

15.2 Capacitance

Capacitors connected in parallel have the same effect as increasing the plate area of one capacitor. Therefore, to calculate the total capacitance, the individual capacitors are added:

$$C_T = C_1 + C_2 + C_3$$

Connecting capacitors in series has the same effect as increasing the thickness of the dielectric. This effectively increases the distance between the plates and reduces the total capacitance of the circuit. The following formula can be used to find the total capacitance of capacitors connected in series:

$$\frac{1}{C_T} = \frac{1}{C_1} + \frac{1}{C_2} + \frac{1}{C_3}$$

The capacitive reactance formula is

$$X_C = \frac{1}{2\pi f C}$$

■ INDUCTANCE

When inductors are connected in series, the total inductance is the sum of the inductances of all the inductors. In other words, inductors behave like resistors:

$$L_T = L_1 + L_2 + L_3$$

In the circuit (Figure 15–2), three conductors are connected in series. L_1 has an inductance value of 0.3 H, L_2 has an inductance value of 0.5 H, and L_3 has an inductance value of 0.6 H. What is the total inductance of the circuit?
Remember,

$$L_T = L_1 + L_2 + L_3$$

Therefore,

$$L_T = 0.3 + 0.5 + 0.6 = 1.4 \text{ H}$$

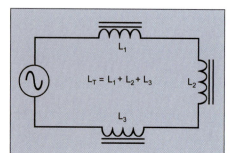

FIGURE 15–2   Inductors in series.

Inductors in parallel behave like resistors in parallel. Therefore, the reciprocal formula can be used:

$$\frac{1}{L_T} = \frac{1}{L_1} + \frac{1}{L_2} + \frac{1}{L_3}$$ (15.1)

15.3 Inductive Reactance

Inductive reactance is the opposition to current flow in an AC circuit caused by the presence of inductance. In an AC circuit that has only inductance, the amount of current that flows is determined by the counterelectromotive force of the single coil.

The induced voltage limits the flow of current through the circuit in a manner similar to resistance. This induced voltage, however, is not resistance, but the effect (inductive reactance) on the circuit limits the current flow in an AC circuit, similar to the effect of resistance. The current-limiting factor of the inductor is called the *reactance*. The symbol for reactance is the letter "X." Since this reactance is caused by the inductance, the symbol X_L is used to represent inductive reactance.

Inductive reactance, like resistance, is measured in ohms and can be calculated when the inductor value and frequency of the source are known. The formula is

$$X_L = 2\pi fL$$

Where:

X_L = inductive reactance

2 = a constant

π = 3.14159

f = frequency in hertz (Hz)

L = inductance in henrys (H).

A convenient substitution that will help you when working with 60-Hz systems is that $2\pi f = 377$. This is much easier to remember than long values of pi (π). Remember that this applies only when $f = 60$ Hz.

The reactance of an inductive circuit depends on the inductance and the frequency of the circuit. Given the frequency and the inductance of a circuit, the inductive reactance can be determined. In order to find the total inductive reactance (X_{LT}) in a series circuit containing more than one inductor, the same method can be used as was used to find inductance. X_{LT} will equal the sum of the inductive reactance for all the inductors:

$$X_{LT} = X_{L_1} + X_{L_2} + X_{L_3}$$

EXAMPLE 1

Three inductors are connected in series. They have inductive reactances of 150 Ω, 220 Ω, and 280 Ω. What is the total inductive reactance of the circuit?

Solution:

$$X_{LT} = X_{L_1} + X_{L_2} + X_{L_3}$$
$$X_{LT} = 150 \ \Omega + 220 \ \Omega + 280 \ \Omega$$
$$X_{LT} = 650 \ \Omega$$

Adding inductors in series with a circuit that already contains inductors will increase the total inductive reactance of the circuit. Also, remember from the previous chapter that inductive reactance is measured in ohms like resistance and can be substituted for R in Ohm's law equations when working with AC circuits.

EXAMPLE 2

A series circuit, made up of three inductors, has a combined inductance of 1.4 H. If the circuit is supplied with a 48-V, 400-Hz source, what is the total circuit current?

Solution:

$$X_L = 2\pi fL$$

$$X_L = 2 \times 3.14159 \times 400 \times 1.4$$
$$X_L = 3518.6 \ \Omega$$

$$I = \frac{E}{X_L}$$

$$I = \frac{48}{3,518.6}$$

$$I = 13.6 \text{ mA}$$

EXAMPLE 3

Now determine what the new current would be if the frequency were dropped to 30 Hz.

Solution:

$$X_L = 2 \times 3.14159 \times 30 \times 1.4$$
$$X_L = 263.89 \ \Omega$$

$$I = \frac{E}{X_L}$$

$$I = \frac{48 \text{ V}}{263.89 \ \Omega}$$

$$I = 182 \text{ mA}$$

In earlier chapters, you learned that in a purely inductive circuit, the current lags the voltage by 90°. Look at Figure 15–3. The total voltage across the inductor voltage is proportional to the rate of change of

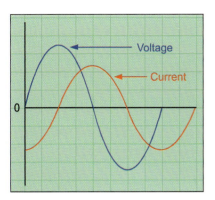

FIGURE 15–3 Current lags voltage in an inductor.

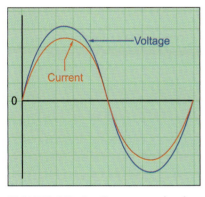

FIGURE 15–4 Current and voltage are in phase in a resistive circuit.

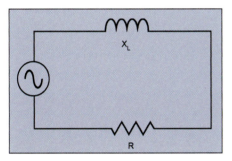

FIGURE 15–5 Schematic of a series RL circuit.

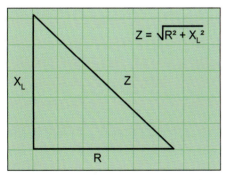

FIGURE 15–6 Phase relationship of a series RL circuit.

the current. At the beginning of the waveform, the current will be at its maximum value in the negative direction. At this time, the current is not changing in magnitude or direction, so the induced voltage is zero. As the current begins to change direction and decrease in value, the magnetic field produced by the flow of current decreases or collapses. The collapsing magnetic field begins to induce voltage into the coil as it is cutting through the conductors.

When an AC voltage is applied to a resistor or purely resistive circuit, the current flow throughout the circuit will be a copy of the voltage. The current and voltage will rise and fall at the same rate and are said to be in phase (see Figure 15–4).

Remember that in an inductive circuit, the induced voltage is able to limit the flow of current through the circuit appearing as resistance. This induced voltage is not resistance. The current-limiting factor is called *inductive reactance*, which is measured in ohms and can be calculated with the values of inductance and frequency because the inductive reactance is proportional to the frequency and the value of the circuit inductors.

15.4 Impedance

In an RL circuit, both resistance and inductive reactance oppose current flow, and their combined opposition is called *impedance*. Resistance is measured in ohms and is represented by the letter "R." Inductive reactance is measured in ohms and is represented by the symbol X_L.

Impedance is also measured in ohms and is represented by the letter "Z." One of the primary rules for series circuits is that the current must be the same in any part of the circuit. Since the total opposition to the current is impedance (Z), it can be used to replace R in the Ohm's law formula.

In series RL circuit (see Figure 15–5), the current is the same throughout the circuit, through both the resistor and the inductor. The voltage dropped across the resistor is in phase with the current, but the voltage dropped across the inductor is 90° out of phase with the current. Therefore, the voltage drops will not be additive.

Using Ohm's law, you can find the voltage drop across the resistor in Figure 15–5: $E_R = I \times R$. To find the voltage drop across the inductor in the circuit, substitute the inductive reactance for resistance in the calculation:

$$E = I \times X_L$$

The relationship between resistance and inductive reactance produces a right triangle (see Figure 15–6). The impedance of the circuit is equal to the length of the hypotenuse. To compute the value of the impedance, the resistance and the reactance must be added vectorially. Since these two components form the legs of a right triangle, the Pythagorean theorem can be used:

$$Z^2 = R^2 + X_L^2 \text{ or } Z = \sqrt{R^2 + X_L^2}$$

EXAMPLE 4

Assume that the coil in Figure 15–5 has an inductive reactance of 100 Ω and that the resistor has a resistance of 150 Ω. What is the total impedance?

Solution:

$$Z = \sqrt{R^2 + X_L^2}$$
$$Z = \sqrt{150^2 + 100^2}$$
$$Z = \sqrt{22,500 + 10,000}$$
$$Z = \sqrt{32,500}$$
$$Z = 180 \ \Omega$$

15.5 Voltage Drop

After the impedance of the circuit is known, it is possible to find the total circuit current I_T. Since the current is the same throughout the series circuit, the individual voltage drops across each component of the circuit can easily be calculated.

EXAMPLE 5

Assume that the source voltage is 45 volts and that R and X_L are the same as before. Calculate the voltage drop for each component.

Solution:
First, find the total current for the circuit:

$$I_T = \frac{E_T}{Z}$$
$$I_T = \frac{45}{180}$$
$$I_T = 0.25 \ A$$

Now calculate the voltage drop across each of the circuit components:

$$E_R = I_R \times R \qquad\qquad E_L = I_L \times X_L$$
$$E_R = 0.25 \times 150 \qquad E_L = 0.25 \times 100$$
$$E_R = 37.5 \ V \qquad\qquad E_L = 25 \ V$$

The voltage drops across the individual components in a series circuit can be added to find the supply voltage. But in looking at the individual voltage drops as calculated in the previous equations, adding the individual voltage drops would indicate that the amount of voltage dropped across the individual components of this series circuit exceeds the source voltage (37.5 V + 25 V = 62.5 V), but the supply volt-

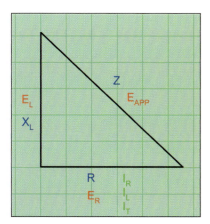

FIGURE 15–7 Phase relationship of all RL circuit components and values.

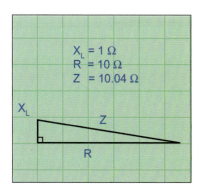

FIGURE 15–8 Large resistance, small inductive reactance.

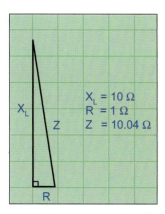

FIGURE 15–9 Small resistance, large inductive reactance.

age is only 45 V. However, the voltage drops across the two components are not in phase and must be added vectorially. Figure 15–7 shows the vector relationships among the various values.

The vector sum of the voltage drops across the resistor and inductor can be found by solving for the resultant, or hypotenuse, of the triangle, using the following formula:

$$E_T = \sqrt{E_R^2 + E_L^2} \qquad E_T = \sqrt{37.5^2 + 25^2} = 45 \text{ V}$$

When a circuit containing both resistance and inductance is connected to alternating current, the total current will lag the applied voltage between 0° and 90°. The exact amount of the phase angle difference is determined by the ratio of resistance as compared to inductance.

If a circuit contains a large amount of resistance and a small amount of inductance, the resistance and impedance will be nearly the same (see Figure 15–8). Also, if there is a large amount of inductance and a small amount of resistance, the impedance and inductive will be nearly the same (see Figure 15–9). It is a ratio. There is a rule of thumb that if the ratio between the reactance and resistance is 10 or greater, the smaller value is considered negligible.

In chapter 9, you learned about the Q of a coil, which is defined as $Q = X_L/R$. When a coil has a large inductive reactance and a small resistance, it has a high Q.

■ POWER FACTOR

The power factor is a ratio of true power to apparent power. The power factor in a series circuit can be computed by using any of the following formulas:

$$PF = \frac{E_r}{E_t}$$

$$PF = \frac{R}{Z}$$

$$PF = \frac{P_T}{P_A} \qquad (15.2)$$

Power factor is generally expressed as a percentage. The decimal fraction must therefore be changed to a percentage by multiplying it by 100.

EXAMPLE 6

Calculate the power factor for the values shown in Figure 15–8.

Solution:

$$PF = \frac{R}{Z}$$

$$PF = \frac{10}{10.04}$$

$$PF = .996, \text{ or } 99.6\% \text{ PF}$$

EXAMPLE 7

Calculate the power factor for the values shown in Figure 15–9.

Solution:

$$PF = \frac{R}{Z}$$

$$PF = \frac{1}{10.04}$$

$$PF = .0996, \text{ or } 9.96\% \text{ PF}$$

15.6 Angle Theta

The phase angle between the voltage and current, or angular displacement, is the angle theta (θ). Since the power factor is the ratio of true power to apparent power, the phase angle of voltage and current is formed between the resistive leg and the hypotenuse. Use the cosine function to find the angle theta for watts divided by the volt-amps. Watts divided by volt-amps is also equal to the power factor.

EXAMPLE 8

Calculate the power factor and the angle theta for the circuit of Figure 15–5. Assume that $X_L = 100 \, \Omega$ and $R = 150 \, \Omega$.

Solution:

Earlier you determined that $Z = 180 \, \Omega$:

$$\frac{R}{Z} = \cos \theta$$

$$\cos \theta = \frac{150}{180} = 0.83333 = PF$$

$$\cos \theta = PF = 0.83333$$

Using a scientific calculator or a table, you find that the angle whose cos is 0.83 is equal to 33.56°:

$$\theta = 33.56°$$

Recall also that

$$\tan \theta = \frac{X_L}{R} = \frac{E_L}{E_R}$$

▮ SERIES RL CIRCUIT RELATIONSHIPS

Remember that all the relationships within an RL circuit are based on a given frequency. As the frequency varies, so do many of the other calculated values. Review each aspect of the circuit relative to the effect that the frequency has on each value as defined by the accompanying formula or equation. Many of the changes are due to acumulative effect of one value changing, which changes another, which changes another, and so on. For example, if frequency is increased in an RL circuit, the following will occur:

■ Impedance will increase because inductance increased:

$$X_\text{L} = 2\pi\, fL$$

■ Current will decrease because impedance increased:

$$I = \frac{E}{Z}$$

■ The circuit will become more inductive:

$$X_\text{L} = 2\pi\, fL$$

■ The voltage across the resistor will decrease because the current through the resistor decreased.

■ The voltage drop across the coil will increase because there is more induced voltage and greater reactance to current flow through the coil.

■ The phase angle will increase because voltage across the resistor dropped.

■ The power factor will decrease because the ratio of the voltage drop across the resistor to the applied voltage became smaller.

■ The resistance will remain the same. The applied voltage will remain the same because the frequency does not affect the power supply, and true power will decrease because the resistive part of the current was decreased, and we know that true power can occur only in the resistive part, not the coil part, of the circuit.

■ SUMMARY

In a pure resistive circuit, the voltage and current are in phase with each other. In a pure inductive circuit, the voltage and current are 90° out of phase with each other. In a series RL circuit, the voltage and current will be out of phase with each other somewhere between 0° and 90°. The ratio between inductive reactance and resistance will determine how out of phase the voltage and current are with each other. Angle theta (θ) is the phase angle difference between the applied voltage and the total circuit current. The cosine of the angle theta is equal to the power factor. The power factor is the ratio of true power to apparent power.

Table 15–1 will be used again in later chapters. It contains both inductive and capacitive relationships.

■ REVIEW QUESTIONS

1. How are the following values calculated in series? In parallel
   a. Resistance
   b. Inductance
   c. Capacitance
   d. Reactance

2. What is the relationship between frequency and inductive reactance? Between frequency and capacitive reactance?

3. Explain in your own words the meaning of "phase angle."

4. What is the range of phase angles for the following series circuits?
   a. Resistance only
   b. Resistance and capacitance
   c. Resistance and inductance
   d. Inductance and capacitance

5. What is power factor? How is it calculated?

Table 15–1 Resistance, Reactance, and Impedance in RL Circuits

	Resistance R, Ω	Inductive Reactance $X_{\mathbf{L}}$, Ω	Capacitive Reactance $X_{\mathbf{C}}$, Ω	Impedance Z, Ω
Definition	In phase opposition to alternating or direct current	opposition to alternating current	opposition to alternating current	Phasor (vector) combination or resistance and reactance $Z = \sqrt{R^2 + X^2}$
Effect of frequency	Same for all frequencies	Increases with higher frequencies	Decreases with higher frequencies	X_L component increases, but X_C decreases
Phase angle θ	0°	I_L lags V_L by 90°	V_C lags I_C by 90°	Series: $\tan \theta = \dfrac{X}{R} = \dfrac{E_X}{E_R}$ Parallel: $\tan \theta = \dfrac{I_X}{I_R}$

■ PRACTICE PROBLEMS

1. Three inductors in series at 60 Hz have values of .3 H, .4 H, and .5 H and have inductive reactances of 188.4 Ω, 251.2 Ω, and 314 Ω. What is the total inductive reactance? Prove your answer.

2. If 100 volts are applied to the circuit in question 1, what is the current?

3. If an X_L of 40 Ω is in series with a resistor of 60 Ω, what is the impedance of the circuit?

4. For the circuit of problem 3, draw the impedance (phasor diagram) triangle including theta. Show your calculations.

5. If 120 volts are applied to the circuit, of problem 3, what is the voltage drop across the resistor and the inductor? Prove your answer.

6. What is the PF of the circuit in problem 3? Show two ways to obtain your answer.

7. Draw the impedance diagram for the following circuit when there is a decrease in frequency to 500 Hz:

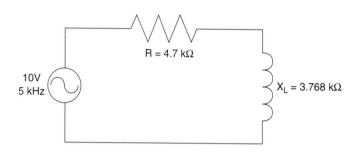

10V
5 kHz

R = 4.7 kΩ

X_L = 3.768 kΩ

chapter **16**

Parallel RL Circuits

■ OUTLINE

■ OVERVIEW

A resistive inductive (RL) circuit contains a resistor and a coil. The previous chapter covered series RL circuits. Just as parallel resistive circuits behave differently than parallel inductive circuits, so do parallel RL circuits behave differently than series RL circuits.

This chapter covers parallel RL circuits in detail and sharpens your skills in circuit analysis for calculating impedances, power factors, and phase angles. An example of the types of circuits covered in this chapter is shown in Figure 16–1.

■ OBJECTIVES

After completing this chapter, you should be able to:

1. Draw circuits with resistors and inductors in parallel.
2. Explain the techniques for solving circuit parameters in parallel RL circuits.
3. Calculate for unknown values in parallel RL circuits.

FIGURE 16–1   A parallel RL circuit.

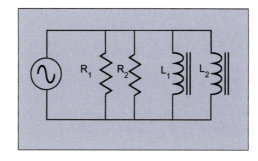

■ RL PARALLEL CIRCUITS

By the nature of its parallel connection, the voltage applied to devices in parallel will be the same for all; consequently, the voltage applied to a resistor and parallel inductor must be in phase and have the same value. The current flow through the inductor, however, will be 90° out of phase (lagging) with the voltage. The current flow through the resistor will be in phase with the voltage. This will produce a phase angle difference of 90° between the current flow of a pure inductive load and the current flow of a pure resistive load. Note that this is the opposite of a series circuit, where the current is the same (common) and the voltage drops are different across each series circuit component (see Figure 16–2).

The phase angle between the total circuit current and the voltage is determined by the ratio of the amount of resistance to the amount of inductive reactance. The circuit power factor is still determined by the ratio of true power to apparent power.

In a series RL circuit, the current is common to all components, and the voltage drops are out of phase. In a parallel RL circuit, the voltage is common to all components, and the currents are out of phase.

■ RL PARALLEL CIRCUIT CURRENT

16.1 Resistive Current

In any parallel circuit, the voltage is the same across each component of the circuit. Therefore, if you know the source voltage or the voltage across any one branch, you know the voltage across all the branches ($E_T = E_1 = E_2 = E_n$). The total current through the resistors, regardless of the number of resistors in parallel, is equal to the sum of the individual currents from each of the parallel branches. With parallel resistors, we say that the currents are additive such that

$$I_T = I_1 + I_2$$

FIGURE 16–2 Resistive and inductive current and voltage phase relationships.

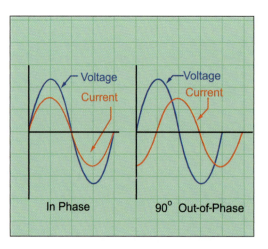

Voltage
Current
In Phase

Voltage
Current
90° Out-of-Phase

The current through any resistive branch of a parallel resistive circuit can be determined utilizing Ohm's law:

$$I_R = \frac{E}{R}$$

16.2 Inductive Current

Just as in the parallel resistive circuit, the current through any branch in a parallel inductor circuit can be found by dividing the applied voltage by the inductive reactance of that branch. The formula being used is a variation of the Ohm's law formula. The amount of current flowing through the parallel inductor or coil can be calculated by using the following formula:

$$\text{First step:} \quad I_L = \frac{E}{X_L}$$

For example, if the voltage is 220 V and the inductive reactance is 20 Ω, what is the current through the coil?

$$\text{Second step:} \quad I_L = \frac{220}{20}$$

$$\text{Third step:} \quad I_L = 11 \text{ A}$$

16.3 Total Current

The total current flow through the circuit is equal to current flow through the resistor added to the current flow through the inductor. Remember that these two parameters are 90° out of phase with each other, so vector addition will be used.

Figure 16–3 shows that the current vectors form a right triangle and that the total current is equal to the length of the hypotenuse. The Pythagorean theorem can be used.

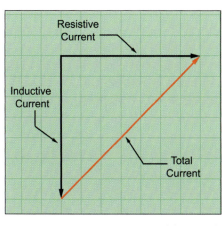

FIGURE 16–3   Vector addition of resistive and inductive currents.

EXAMPLE 1

In Figure 16–1, assume that the resistive current is 15 A and that the inductive current is 12 A. Calculate the total current flow in the circuit.

Solution:

$$I_T^2 = I_R^2 + I_L^2$$
$$I_T^2 = 15^2 + 12^2$$
$$I_T = \sqrt{369} = 19.2 \text{ A}$$

■ IMPEDANCE

The impedance in a parallel circuit having both resistance and inductance is the total opposition to the flow of current. That total opposition to the flow of current is the result of the opposition to the current

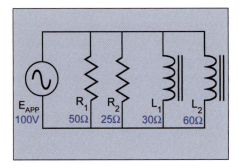

FIGURE 16–4   Multiple resistors and inductors in parallel.

flow from the resistive branch added vectorially to the opposition to the current flow from the inductive branch. Since the relationship of R and the X_L are vector quantities, they must be added vectorially. The formula used for that vector addition is

$$Z = \frac{R \times X_L}{\sqrt{R^2 + X_L^2}}$$

When a circuit has more than one resistor or more than one inductor, the values used in the previous problem represent the total amount of resistance from all the resistors and the total amount of inductive reactance from all the inductive reactances.

EXAMPLE 2

Calculate the total impedance for the circuit shown in Figure 16–4.

Solution:
First, find R_T and X_{LT}:

$$R_T = \frac{50 \times 25}{50 + 25} = 16.6 \ \pi\Omega$$

$$X_{LT} = \frac{30 \times 60}{30 + 60} = 20 \ \pi\Omega$$

First method—product over the sum (magnitude only):

$$Z = \frac{R \times X_L}{\sqrt{R^2 + X_L^2}}$$

$$Z = \frac{16.6 \times 20}{\sqrt{16.6^2 + 20^2}} = 12.8 \ \Omega$$

Second method—product over the sum (vector solution):

$$Z_T = \frac{16.6\angle 0° \times 20\angle 90}{\sqrt{(16.6\angle 0°)^2 + (20\angle 90°)^2}}$$

$$Z_T = \frac{(16.6 \times 20)\angle(0° + 90°)}{(16.6\angle 0°) + (20\angle 90°)}$$

Applying a scientific calculator to perform the vector calculations yields

$$Z_T = 12.8\angle 39.69° \ \Omega$$

Note that the impedance value is smaller than either of the total resistance values or the total inductive reactance values calculated for the circuit ($R_T = 16.6 \ \Omega$, $X_{L \ tot} = 20 \ \Omega$). Had the total circuit current been known, the following Ohm's law formula variation could have been used:

$$Z = \frac{E}{I_T}$$

EXAMPLE 3

Assume that the circuit in Figure 16–4 had a total current of 7.81 A. Calculate the circuit impedance:

Solution:

$$Z = \frac{100}{7.81}$$

$$Z = 12.8 \ \Omega$$

As can be seen, the value for the impedance, calculated by both methods, is the same.

■ POWER CALCULATIONS

16.4 Apparent Power

Apparent power can be calculated by finding the product of the total current flow and the circuit voltage. The relationship for volt-amps, watts, and volts-amps-reactive (VARs) is the same for a parallel RL circuit as it was for a series RL circuit. Remember that the total power is equal to the sum of the individual powers in any type of circuit. Look at Figure 16–5 for an example. The power of branch 1 + branch 2 + branch 3 + branch 4 will yield total circuit power. Since true power and reactive power are 90° out of phase with each other, they form a right triangle with apparent power as the hypotenuse.

FIGURE 16–5 Apparent power, true power, and reactive power.

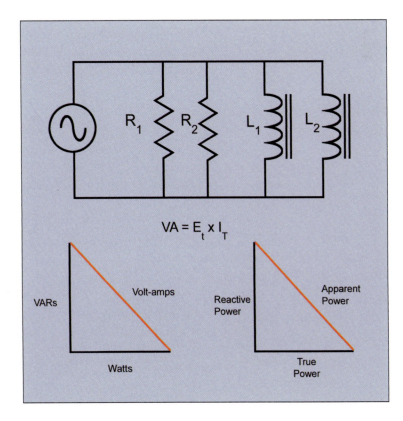

True power in an RL circuit is consumed by the resistance in the circuit. The reactive power is returned each ¼ cycle to the source by the collapsing magnetic field around the coil. In both a series and a parallel RL circuit, the power delivered by the source is the apparent power. If the frequency of the parallel RL circuit is increased, the inductive reactance in the circuit will increase, so the total circuit impedance will increase:

$$X_L = 2\pi fL$$

Therefore,

$$Z = \frac{R \times X_L}{\sqrt{R^2 + X_L^2}}$$

16.5 Power Factor

Power factor (PF) in a parallel RL circuit is the ratio of true power to apparent power, as it is in any AC circuit. The formula for calculating PF in a parallel RL circuit is

$$PF = \frac{\text{true power}}{\text{apparent power}}$$

In terms of voltage and current, this formula can be further defined:

$$PF = \frac{I_R \times E_{APP}}{I_{tot} \times E_{APP}} = \frac{I_R}{I_{tot}}$$

As can be seen from this formula, since the applied voltage, E_{APP}, is part of the true power and apparent power calculation (because the voltage is the same in all parts of a parallel circuit), the E_{APP} values will cancel each other out. The resulting formula uses just the current through the resistor and the total current to calculate the power factor.

16.6 Angle Theta

The angle theta is the angle produced by drawing the triangle representing the vector relationship of the true power, apparent power, and reactive power (see Figure 16–6). The cosine of the angle theta, the angle formed by the true power–apparent power relationship, is equal to the power factor. Once the values for the true power and apparent power are known, finding the power factor can be done as follows:

$$PF = \cos \theta = \frac{\text{true power}}{\text{apparent power}}$$

From this example, you can see that if a circuit has a true power of 12 watts and an apparent power of 15 volt-amps, the calculation for the power factor and the phase angle is

$$PF = \frac{\text{true power}}{\text{apparent power}}$$

$$PF = \frac{12 \text{ watts}}{15 \text{ volt-amps}}$$

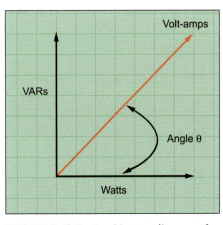

FIGURE 16–6   Vector diagram for angle theta.

$$PF = 0.8, \text{ or } 80\% \text{ PF}$$

$$PF = \cos \theta = \frac{\text{true power}}{\text{apparent power}}$$

$$\cos \theta = .8$$

$$\theta = 36.87°$$

■ SUMMARY

The voltage applied across parts of a parallel circuit must be the same. The current flowing through the resistive parts of a parallel RL circuit will be in phase with the voltage. The current flowing through the inductive parts of the circuit will lag the voltage by 90°. The total current through the parallel circuit is equal to the sum of the individual currents. Vector addition must be used since the current through the inductive parts is 90° out of phase with the current flow through the resistive parts:

$$I_T = \sqrt{I_R^2 + I_L^2}$$

The impedence in a parallel RL circuit can also be computed by using vector addition to add the reciprocals of the resistance and inductive reactance.

Apparent power is the sum of true power and reactive power in any kind of circuit. Vector addition must also be utilized to add these values since true power and reactive power are 90° out of phase with each other.

■ REVIEW QUESTIONS

1. How do the total voltage and current in a parallel RL circuit compare to the resistive current and voltage? Inductive current and voltage?

2. Using the resistance and reactance of a parallel RL circuit, show the different methods for calculating the total impedance.

3. Discuss the relationships among the following in a parallel circuit:
   a. Total apparent power
   b. Total true power
   c. Total reactive power
   d. Power

■ PRACTICE PROBLEMS

All practice problems refer to the following figure:

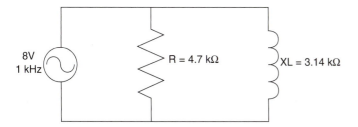

1. What is the current through the resistive branch?

2. What is the current through the inductive branch?

3. Find I_T.

4. Construct a current (phasor diagram) triangle. Note that voltage is now the reference vector.

5. What is the impedance of this circuit?

6. Show how to find the impedance in problem 5 using the parallel impedance formula.

7. Find the PF in problem 1.

8. A decrease in frequency in a parallel circuit will cause the PF to _____.

chapter 17

Series RC Circuits

■ OVERVIEW

All series circuits have two principles in common:

1. The current in all elements of the series circuit is equal to the total current.
2. The sum of the voltage drops across all circuit elements (including the sources) is equal to zero.

Series RL circuits were covered in chapter 15. In addition to the two principles in common to all series circuits, the voltage drop across the resistor is in phase with the current, and the voltage drop across the inductor leads the current by 90°.

A circuit that contains resistance and capacitance is called an *RC circuit*. You will find that the series RC circuit is similar to the RL circuit in that it uses the current as the reference. The only difference is that the voltage drop across the capacitor lags the current by 90° (see Figure 17–1). This chapter uses the same principles you learned previously to analyze RL circuits.

■ OBJECTIVES

After completing this chapter, you should be able to:

1. Explain the relationships that exist between parameters in a series RC circuit.
2. Calculate mathematically for various unknown values in a series RC circuit.

FIGURE 17–1 RC and RL current–voltage relationships.

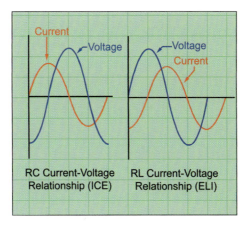

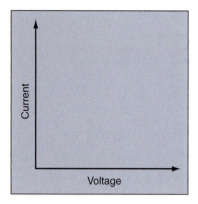

FIGURE 17–2 Capacitive circuit voltage–current relationship.

RESISTIVE CAPACITIVE SERIES CIRCUITS

If a pure capacitive load is connected in an AC circuit, the voltage and current are 90° out of phase with each other. In a capacitive circuit, the current leads the voltage by 90° (see Figure 17–2). Notice that the voltage is always drawn as the reference at 0°C.

When a circuit that contains both resistance and capacitance is connected to an AC power supply, the voltage and current will be out of phase by some amount between 0° and 90°. The size of the phase angle will be determined by the ratio between resistance and capacitive reactance.

Resistive capacitive series circuits are similar to resistive inductive series circuits previously covered. The formulas are basically the same with only minor modifications. Figure 17–3 shows a 20-Ω resistor in series with a capacitor having 30 Ω of capacitive reactance. The voltage supply is 220 V at 60 Hz. This circuit will be used for several of the examples in this chapter.

17.1 Impedance

Remember that total impedance is equal to the total opposition to current flow in the circuit. It is a combination of both resistance and capacitive reactance. Since this is a series circuit, the elements are added. Resistance and capacitive reactance are 90° out of phase with each other, and this forms a right triangle with the impedance forming the hypotenuse (see Figure 17–4). The formula for Z is

$$Z = \sqrt{R^2 + X_C^2}$$
$$Z = \sqrt{400 + 900}$$
$$Z = 36\ \Omega \qquad (17.1)$$

17.2 Capacitive Reactance Phase Angle

Notice in Figure 17–2 that the current *leads* the voltage by 90°; however, in Figure 17–4, the capacitive reactance *lags* the resistance by 90°.

FIGURE 17–3 Series RC circuit.

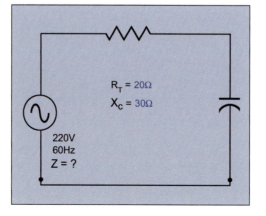

$R_T = 20\Omega$
$X_C = 30\Omega$

220V
60Hz
Z = ?

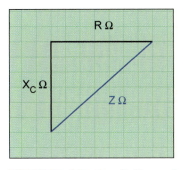

FIGURE 17–4 *R*, X_C, and impedance (*Z*).

To understand why this happens, recall the vector values for voltage and current in Figure 17–2:[1]

$$\overline{E}_C = E_C\angle 0° \; and \; \overline{I}_C = I_C\angle +90° \tag{17.2}$$

From Ohm's law,

$$X_C = \frac{E_C}{I_C} = \frac{E_C\angle 0°}{I_C\angle 90°} \tag{17.3}$$

Remember from your vector theory that to divide two vectors, you divide their magnitudes (in this case, E_C and I_C) and subtract the angles (in this case 0° and 90°). This means that

$$X_C = \frac{E_C}{I_C}\angle(0° - 90°) = \frac{E_C}{I_C}\angle -90° \tag{17.4}$$

17.3 Total Current

With total impedance known, the total current of the circuit can be determined:

$$I = \frac{E}{Z}$$
$$I = \frac{220}{36}$$
$$I = 6.1 \text{ A}$$

17.4 Voltage Drop Across the Resistor

In series circuits, remember that the current is the same at any point in the circuit. For example, 6.1 A of current flow will be through both the resistors and the capacitors. The voltage drop across the resistors (E_R) can be computed by using the following formula:

$$E_R = I \times R$$
$$E_R = 6.1 \times 20$$
$$E_R = 122 \text{ V}$$

17.5 True Power

True power for the circuit is calculated by using any of the power formulas as long as they are values that apply to the resistive part of the circuit (remember that only resistive components produce watts). Also remember that current and voltage must be in phase with each other for true power to exist:

$$P = E_R \times I$$
$$P = 122 \times 6.1$$
$$P = 744.2 \text{ W}$$

---

[1]Any value with a bar over the top, such as $\overline{E}_C$, indicates the vector value.

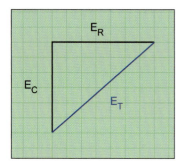

FIGURE 17–5   Voltage drops across the resistor and the capacitor form a right triangle.

17.6 Capacitance

Remember that capacitance is calculated using the following formula:

$$C = \frac{1}{2\pi f X_C}$$

$$C = \frac{1}{6.28 \times 60 \times 30}$$

$$C = \frac{1}{11,034}$$

$$C = 0.000088464 \text{ F, or } 88.464 \text{ μF}$$

17.7 Voltage Drop Across the Capacitor

The voltage drop across the capacitors (E_C) is calculated using the following formula:

$$E_C = I \times X_C$$

$$E_C = 6.1 \times 30$$

$$E_C = 183 \text{ V}$$

17.8 Total Voltage

As long as the current is known, total voltage can be calculated. Remember that in a series circuit, the voltage drop across the capacitor and the resistor are 90° out of phase with each other, so you must use vector addition. The two voltage drops form the legs of the right triangle (see Figure 17–5).

Total voltage can be calculated using the following formula:

$$E_T = \sqrt{E_R^2 + E_C^2}$$

$$E_T = \sqrt{122^2 + 183^2}$$

$$E_T = \sqrt{14,884 + 33,489}$$

$$E_T = \sqrt{48,373}$$

$$E_T = 220 \text{ V}$$

17.9 Apparent Power

The apparent power of the circuit is calculated by using the total values of voltage and current. The following formula is used:

$$VA = E_T \times I$$

$$VA = 220 \times 6.1$$

$$VA = 1,342$$

17.10 Power Factor

The power factor is the ratio of true power to apparent power. Since the values of true power and apparent power have been calculated previously, those values can be used to calculate the power factor of the circuit as follows:

$$PF = \frac{\text{true power}}{\text{apparent power}}$$

$$PF = \frac{\text{watts}}{\text{volt-amps}}$$

$$PF = -\frac{744.2 \text{ W}}{1342 \text{ VA}}$$

$$PF = 0.5545, \text{ or } 55.45\%$$

The power factor can also be calculated in a series circuit using the relationship between the source voltage E_T and the voltage across the resistor E_R. While the values will be different than those used for power, the ratio of those values will be the same. As long as the phase angle is the same, the power factor will also be the same. Using the values from Figure 17–3, the power factor is calculated as

$$PF = \frac{E_R}{E_T}$$

$$PF = \frac{122}{220} = 0.5545, \text{ or } 55.45\%$$

17.11 Angle Theta

The power factor of the circuit is the cosine of the phase angle. Since the power factor turned out to be 0.5545, or 55.45%, angle theta ($\angle \theta$) can be calculated:

$$\cos \theta = PF$$

$$\cos \theta = 0.5545$$

$$\theta = 56.32°$$

In this circuit, the current leads the applied voltage by 56.32°.

■ SUMMARY

In a pure capacitive circuit, the voltage and current are 90° out of phase with each other. In a pure resistive circuit, the voltage and current are in phase with each other. In a circuit containing both resistance and capacitance, the voltage and current will be out of phase with each other by some angle between 0° and 90°.

The amount of phase angle difference between voltage and current in a series RC circuit is equal to the angle whose tangent is the ratio of resistance to capacitive reactance.

In a series circuit, the current flow is the same through all components of the circuit. True power can be found only in the resistive parts of the circuit. The power factor is the ratio of true power to apparent power.

■ REVIEW QUESTIONS

1. Compare the following between an RL circuit and an RC circuit:

   a. Phase angle

   b. Power factor

   c. Resistive voltage drop

   d. Reactance voltage drop

2. Discuss impedance in a series RC circuit.

   a. How can it be calculated?

   b. How will it compare to the resistance and the capacitance?

   c. How is it calculated?

3. In your own words, explain why the capacitive reactance lags the resistance by 90°, while the capacitive current leads the capacitive current by 90°.

■ PRACTICE PROBLEMS

1. Current _____ voltage in an RC circuit.

2. The reference vector in a series circuit is generally the _____ vector on the horizontal axis.

3. Draw and label an impedance (phasor diagram) triangle for a series RC circuit.

4. In the following figure, if E_C = 80 volts, what is the voltage across the resistor?

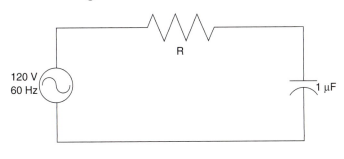

5. What is the X_C in problem 4?

6. What is I_T?

7. What is the value of the resistor in problem 3?

8. Find the impedance of the circuit of problem 3. Prove it.

9. Find PF.

10. A circuit has E_T = 15 V at 2 KHz with a resistor (3 KΩ) and a capacitor in series. If 4 mA flow, what is the value of the capacitor?

chapter **18**

Parallel RC Circuits

■ **OUTLINE**

■ OVERVIEW

In the previous chapter, the series RC circuit was discussed. Remember that in series circuits, current is the same and is in phase through all the components. In series circuits that contain both resistance and reactance, the total voltage and the capacitive voltage are out of phase with the current.

In a parallel circuit, voltage is the same across all the components. The voltage is the same and is in phase. In parallel circuits that contain both resistance and reactance, the currents are out of phase, and the voltage is the reference. Again, a parallel circuit that contains resistance and capacitance is called an RC circuit.

When solving for the power factor or phase angle ($\angle \theta$) in series circuits, voltage drops or resistance and reactance values are used; in parallel circuits, branch currents or resistance and reactance values are used.

■ OBJECTIVES

After completing this chapter, you should be able to:

1. Describe the effects of changing parameters in a parallel RC circuit.
2. Determine mathematically unknown quantities in these circuits.

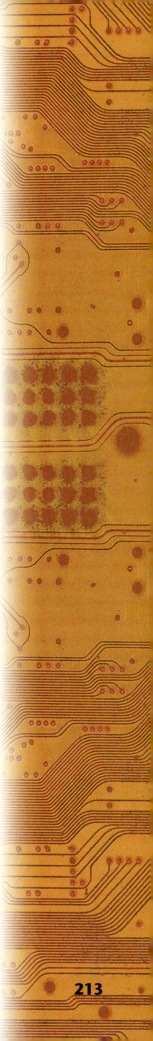

■ RESISTANCE AND CAPACITANCE REVIEW

When resistors are connected in parallel, there are several ways to calculate total resistance. When all the resistors are of equal value, the total resistance is equal to the resistance of one of the resistors, or branches, divided by the number of resistors, or branches (N):

$$R_T = \frac{R}{N}$$

If the resistors are not equal, the reciprocal formula can be used:

$$R_T = \frac{1}{\dfrac{1}{R_1} + \dfrac{1}{R_2} + \dfrac{1}{R_3}}$$

A method for the total resistance of two resistors in parallel is to divide the product of resistances by their sum. The process can be repeated until the last pair has been combined. This is called the *product-over-sum method* for finding total resistance:

$$R_T = \frac{R_1 \times R_2}{R_1 + R_2}$$

Connecting capacitors in parallel has the same effect as increasing the plate area of one capacitor. Capacitors in parallel are calculated similar to inductors and resistors in series.

The total capacitance of a parallel capacitance circuit is given by

$$C_T = C_1 + C_2 + C_3$$

EXAMPLE 1

Calculate the total capacitance of the circuit shown in Figure 18–1.

Solution:

$$C_T = C_1 + C_2 + C_3$$
$$C_T = 20 + 30 + 60$$
$$C_T = 110 \ \mu F$$

FIGURE 18–1 Capacitors connected in parallel.

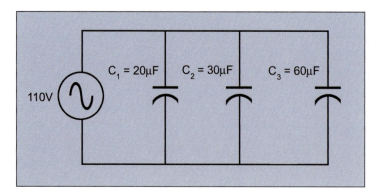

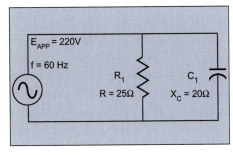

FIGURE 18–2   Parallel RC circuit.

RC PARALLEL CIRCUITS

When resistance and capacitance are connected in parallel, the voltage across all the devices will be in phase and will have the same value. The current flow through the capacitor will be 90° out of phase with the current flow through the resistor. The phase angle shift between the total current and total voltage is determined by the ratio of the amount of resistance to the amount of capacitive reactance. Since the amount of capacitive reactance and amount of resistance will independently determine the amount of current flowing through each branch, whether it is the resistive or capacitive branch, the total current flow can be determined only after those values are determined. The capacitive branch of the circuit is directly affected by the frequency of the source.

18.1 Circuit Values

In Figure 18–2, the resistor (25 Ω) is connected in parallel with a capacitor that has a capacitive reactance of 20 Ω. The circuit power supply is 220 V at a frequency of 60 Hz. Remember that the resistive and capacitance parts will have to be worked separately.

18.2 Resistive Current

The current flow through the resistor can be calculated using the following formula:

$$I_R = \frac{E}{R}$$

$$I_R = \frac{220}{25}$$

$$I_R = 8.8 \text{ A}$$

18.3 True Power

Remember that true power is in the resistive parts of a circuit where current flow is in phase with the voltage. The amount of true power in the circuit can be determined by using any of the values associated with the purely resistive part of the circuit. For Figure 18–2, true power can be found using the following formula:

$$P = E \times I_R$$

$$P = 220 \times 8.8$$

$$P = 1{,}936 \text{ W}$$

18.4 Capacitive Current

Remember that for capacitive parts of circuits, the capacitive reactance can be substituted for resistance in Ohm's law. The current flow through the capacitor can be calculated using the following formula:

$$I_C = \frac{E}{X_C}$$

$$I_C = \frac{220}{20}$$

$$I_C = 11 \text{ A}$$

The capacitance of the capacitor can be calculated using the following formula:

$$C = \frac{1}{2\pi f X_C}$$

$$C = \frac{1}{6.28 \times 60 \times 20}$$

$$C = \frac{1}{7,536}$$

$$C = .000133 \text{ F, or } 133 \text{ } \mu\text{F}$$

18.5 Total Current

Remember that the voltage across all legs of a parallel circuit is the same. The current flow through the resistive leg is in phase with the voltage, and the current flow through the capacitor is leading the voltage by 90° (see Figure 18–3).

The 90° difference in resistive and capacitive current forms a right triangle. Since the two currents are connected in parallel, vector addition must be used to find the total current flow for the circuit. The total current flow can be calculated using the following formula (see Figure 18–4):

$$I_T = \sqrt{I_R^2 + I_C^2}$$

$$I_T = \sqrt{8.8^2 + 11^2}$$

$$I_T = \sqrt{77.44 + 121} = 14.09 \text{ A}$$

FIGURE 18–3 Current leads voltage by 90° in the capacitive circuit.

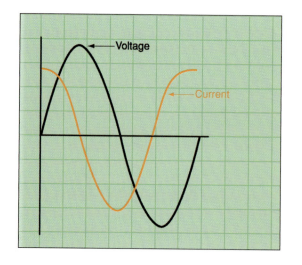

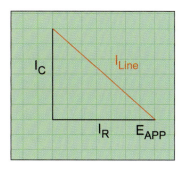

FIGURE 18–4 Parallel RC voltage–current relationship.

18.6 Impedance

The total circuit impedance can be found using any of the total values and substituting Z for R in the Ohm's law formula. The total impedance can be calculated using the formula

$$Z = \frac{E}{I_T}$$

$$Z = \frac{220}{14}$$

$$Z = 15.7 \ \Omega$$

The impedance can also be found by using any of the various formulas. Remember that since the values are not in phase, vector calculations must be used:

$$Z = \frac{R \times X_C}{R + X_C} = \frac{(25\angle 0) \times (20\angle -90)}{(25\angle 0) + (20\angle -90)} \tag{18.1}$$

Using a scientific calculator to perform the operations yields

$$Z = 15.62\angle -51.3 \tag{18.2}$$

The slight difference is caused by rounding errors.

■ APPARENT POWER

The apparent power can be computed by multiplying the circuit voltage times the total current flow:

$$VA = E \times I_T$$
$$VA = 220 \times 14$$
$$VA = 3{,}080$$

18.7 Power Factor

The power factor is the ratio of true power to apparent power. The formula for determining the power factor is the following:

$$PF = \frac{\text{true power}}{\text{apparent power}}$$

$$PF = \frac{W}{VA} \times 100$$

$$PF = \frac{1{,}936}{3{,}080} \times 100$$

$$PF = 62.9\%$$

18.8 Angle Theta

Remember that the cosine of angle theta is equal to the power factor:

$$\cos \theta = 0.628$$
$$\theta = 51°$$

■ SUMMARY

The current flow in the resistive part of the circuit is in phase with the voltage. True power can be calculated by utilizing the resistive current and the applied voltage. The current flow in the capacitor part of the circuit leads the voltage by 90°.

The amount that the total current and the applied voltage are out of phase with each other is deter-

mined by the ratio of resistance to capacitive reactance. The voltage is the same across any leg of a parallel circuit. The circuit power factor is the ratio of true power to reactive power.

■ REVIEW QUESTIONS

1. Compare the following between an RL circuit and an RC circuit:
   a. Phase angle
   b. Power factor
   c. Resistive voltage drop
   d. Reactance voltage drop
2. Discuss impedance in a series RC circuit.
   a. How can it be calculated?

   b. How will it compare to the resistance and the capacitance?
   c. How is it calculated?
3. In your own words, explain why the capacitive reactance lags the resistance by 90°, while the capacitive current leads the capacitive current by 90°.

■ PRACTICE PROBLEMS

1. A 5-V circuit at 100 KHz has a 1,000-pF capacitor in parallel with a 1,000-Ω resistor. Find I_R.

2. Find X_C.

3. Find I_C.

4. Find I_T.

5. Find Z.

6. Find the phase angle.

7. Find PF.

8. Prove the answer to problem 5 using $Z = RX_C/\sqrt{R^2 + X_C^2}$.

9. What is the total power delivered to this circuit?

10. What is the true power in this circuit?

11. In the same circuit, will an increase in frequency to 200,000 Hz cause this circuit to be more or less capacitive? Prove your answer.

chapter 19

Inductive and Capacitive (LC) Circuits

■ OUTLINE

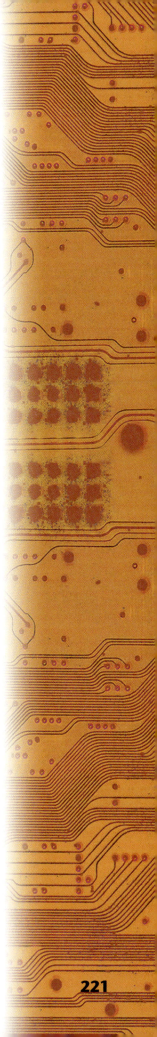

■ OVERVIEW

This chapter discusses both series and parallel LC circuits. When only inductance and capacitance are included, some very interesting conditions result. LC circuits are at the heart of many electrical and electronic applications, such as radio, television, and some control applications.

■ OBJECTIVES

After completing this chapter, you should be able to:

1. Identify series and parallel LC circuits.
2. Describe how circuit parameters change when various values are given.
3. Calculate for unknown values in LC circuits.

■ GLOSSARY

Resonant frequency The frequency at which the inductive reactance and the capacitive reactance are equal in magnitude.

■ SERIES LC CIRCUITS

The voltage across an inductor leads the current by 90°, and the voltage across the capacitor lags the current by 90°. Using the current as the reference yields the following Ohm's law formulas:

$$\overline{X_C} = \frac{E_C \angle -90}{I} \qquad (19.1)$$

and

$$\overline{X_L} = \frac{E_L \angle 90}{I} \qquad (19.2)$$

Since the current in a series circuit is the same in both cases, you can conclude that in a series inductive capacitive (LC) circuit, X_L and X_C are 180° out of phase with each other. Therefore, the value of one will subtract from the other, and the resulting remainder will determine if the circuit is inductive or capacitive (see Figure 19–1).

In Figure 19–1, the current is used as the reference for clarity. Remember that many of the diagrams that you will see in your career as an electrician will use the voltage as the reference.

■ PARALLEL LC CIRCUITS

In a parallel circuit, voltage will be the same across each component, or leg, and is used as the reference, as shown in Figure 19–2. Here I_L and I_C are 180° out of phase, with the larger one determining whether the circuit is inductive or capacitive.

■ SERIES RLC CIRCUITS

When an AC circuit contains the elements of resistance, inductance, and capacitance connected in series, the current is the same through

FIGURE 19–1 LC combination series circuit.

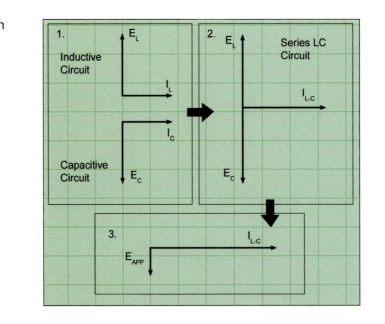

FIGURE 19–2   LC combination parallel circuit.

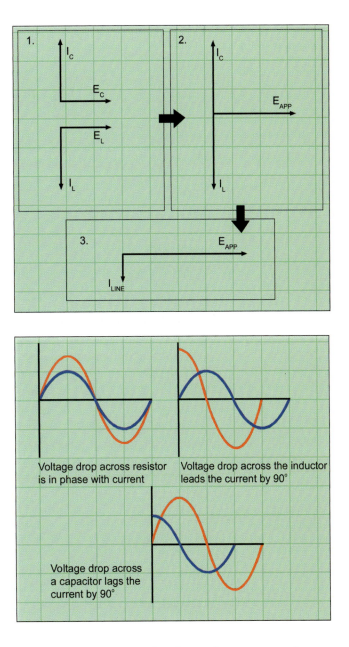

FIGURE 19–3   Voltage and current relationships in a series RLC circuit.

Voltage drop across resistor is in phase with current

Voltage drop across the inductor leads the current by 90°

Voltage drop across a capacitor lags the current by 90°

all parts, but the total voltage drop across the components is out of phase with the current. This is a characteristic of an RLC series circuit.

The voltage drop across the resistor will be in phase with the current, which means that true power is dissipated by the resistor. The voltage drop across the inductor will lead the current by 90°, and the voltage drop across the capacitor will lag the current by 90°. These phase relationships can be seen in Figure 19–3.

The ratio of total reactance to resistance will determine how much the applied voltage will lag or lead the circuit current. If the circuit is more inductive than capacitive, that current will lag the applied voltage, and the power factor will be a lagging power factor. If the circuit is more capacitive than inductive, the current will lead the voltage, and the power factor will be a leading power factor.

Since inductive reactance and capacitive reactance are 180° out of phase with each other, they cancel each other in an AC circuit. This cancellation means that the total reactance is the difference between the two

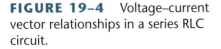

FIGURE 19–4 Voltage–current vector relationships in a series RLC circuit.

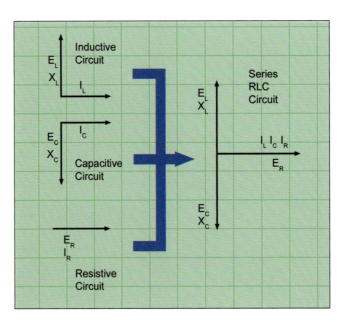

types of reactance. The total reactance could actually be less than either or both of the reactances, allowing a high amount of current through the circuit. Figure 19–4 shows the composite vector representation of the individual RLC circuit components as they relate to each other.

When Ohm's law is used on the circuit values, the voltage drops developed across the reactive components can be higher than the applied voltage. This may seem counterintuitive; however, remember that the inductive voltage and the capacitive voltage will subtract from each other, so Kirchhoff's voltage law still works.

■ PARALLEL RLC CIRCUITS

When an AC circuit has the elements of resistance, inductance, and capacitance connected in parallel, the rules associated with a parallel circuit change the function of the components. Remember that the voltage drop across each leg or component is the same. The currents flowing through each branch, or leg, will be out of phase with each other.

Take a look at Figure 19–3. The current flowing through the resistive component will be in phase with the applied voltage. The current flowing through the inductive part will lag the applied voltage by 90°, and the current flowing through the capacitive leg will lead the applied voltage by 90°. The phase angle difference between the applied voltage and the total current is determined by the ratio of resistance to total reactance connected in parallel.

Figure 19–5 shows the vector relationship among the components in a parallel RLC circuit. In a series RLC circuit, the larger of the two reactances determines whether the circuit is inductive or capacitive. If the inductive reactance is larger, the circuit is inductive; however, if the capacitive reactance is larger, the circuit is capacitive.

The opposite is true in a parallel RLC circuit. If X_L is greater, more current flows in the capacitive branch, and the net reactance is capacitive. If X_L is smaller, the circuit will be inductive in nature because more inductive current will flow.

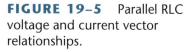

FIGURE 19–5 Parallel RLC voltage and current vector relationships.

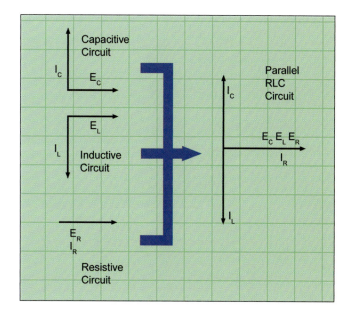

■ SERIES AND PARALLEL LC CIRCUITS

Removing the resistance from the circuit leads to some very interesting results. Although the resistance can never be completely removed, the wires that make up capacitors and inductors do not have much resistance compared to the inductance and capacitance of the units. Therefore, a capacitor and an inductor only in a circuit will behave very much as if there is no resistance present.

Remember that in a series circuit, the current will be the same at all points in the circuit. The voltage drop across each of the elements (a capacitor and an inductor) will be determined by the reactance of each element. The two voltage drops will be 180° out of phase with each other. Figure 19–4 illustrates this condition vectorially.

To further explain this, use a familiar word problem.

EXAMPLE 1

A train leaves town and travels west for 100 miles, then reverses and travels due east for 40 miles on the same track. How far is the train from its starting point?

Solution:
Think of the two distances as though they were two vectors—one vector is 100 miles west, and the other is a vector 40 miles east. These can be represented by vector notation as

   $100 \angle 180°$ and $40 \angle 0°$

Where:

   $180° =$ west

   $0° =$ east.

FIGURE 19–6   Diagram for
Example 1.

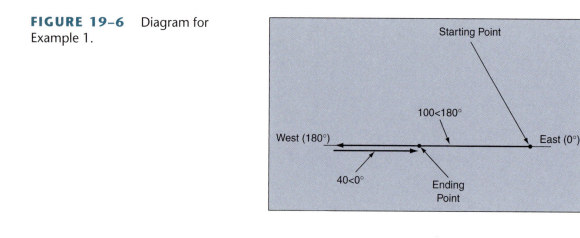

FIGURE 19–6   Diagram for
Example 1.

FIGURE 19–7   Sample series LC
circuit.

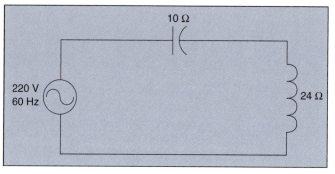

Figure 19–6 shows the calculations. Notice that Example 1 could be changed to north and south by using angles of 90° and −90° instead of 180° and 0°.

19.1 Series LC Circuits: Total Impedance

Since inductive reactance and capacitive reactance are 180° out of phase with each other, vector addition must be used to find their sum. This method results in the smaller of the two being subtracted from the larger. The smaller value is eliminated, and the larger is reduced by the amount of the smaller value.

The circuit of Figure 19–7 is inductive because 24 − 10 = 14 (of the remaining inductive reactance). Figure 19–8 shows this vectorially. Note that this is the electrical equivalent of the train example worked earlier. Instead of east and west, the vectors are going north and south; however, the result is the same.

The current through the circuit can be calculated using Ohm's law:

$$I_T = \frac{E}{Z} = \frac{220\ V}{14\ \Omega} = 15.7\ \Omega \tag{19.3}$$

The voltage drop across each component can be calculated using Ohm's law. Remember that the current through each element is the same in a series circuit:

$$E_L = I \times X_L = 15.7A \times 24\ \Omega = 377.14\ V$$
$$E_C = I \times X_C = 15.7A \times 10\ \Omega = 157\ V \tag{19.4}$$

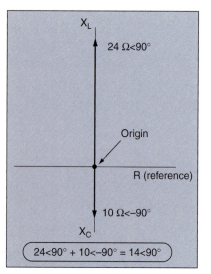

FIGURE 19–8   Calculation of a
series LC circuit.

Notice that the two voltages are 180° out of phase so that the total voltage can be calculated by subtracting one from the other:

$$E_T = E_L - E_C = 377.14 - 157 = 220.14 \text{ V} \tag{19.5}$$

As in previous examples, the slight difference is caused by calculation rounding.

19.2 Parallel LC Circuits: Total Impedance

Method 1

If the two components are connected in parallel (see Figure 19–9), it will be capacitive. In the series connection of Figure 19–8, if X_L is increased, its voltage drop increases. This means that the circuit becomes more inductive. If X_L is larger than X_C, the circuit is inductive. If X_C is larger, the circuit is capacitive.

In a parallel circuit, the voltage is the same across all branches. As the X_L or the X_C changes, the current through that component changes. As X_L goes up, the current through the inductor will go down. This means that the total circuit current, which is the sum of the capacitive and inductive current, will become less inductive. The component (inductor or capacitor) with the smaller reactance will determine whether the circuit is inductive or capacitive in nature.

The relationship between the reactance and the current is an inverse relationship. As the reactance goes up, the current goes down. Remember that the current through the inductive branch is 180° out of phase with the current through the capacitive branch. The resultant can be found by subtracting the smaller current from the larger current. The resultant current will determine whether the circuit is inductive or capacitive. Another method used to determine if a circuit is capacitive or inductive is the reciprocal method. In Figure 19–9, ¹⁄₁₀ is greater than ¹⁄₂₄, which means the circuit is capacitive.

Since the circuit is capacitive, the next step is to determine the impedance using a somewhat modified parallel formula:

$$X_T = \frac{X_L \times X_C}{X_L - X_C} \tag{19.6}$$

$$X_T = \frac{24 \times 10}{24 - 10} = 17.14 \ \Omega \tag{19.7}$$

FIGURE 19–9 Sample parallel LC circuit.

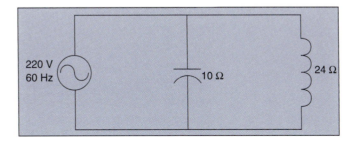

Two important points must be considered:

1. The term X_T is used instead of Z. Although either could be used, reactance is used since there is no resistance in this particular circuit.
2. The minus sign instead of a plus sign in the denominator is necessary because the inductive reactance and the capacitive reactance are opposite each other. Remember the railroad train example discussed previously.

Since the X_C is smaller than the X_L, allowing more current to flow through the capacitive branch of the parallel circuit, the parallel circuit is capacitive. If the circuit were inductive in nature, having an X_L value smaller than the X_C value and allowing more current to flow through the inductive branch, the following formula would be used:

$$X_T = \frac{X_L \times X_C}{X_C - X_L} \tag{19.8}$$

Note that the only difference is the change in the order of the values in the denominator, where the order is arranged to allow for the answer to be a positive value.

Method 2

A better way, although perhaps a little more complex, is to actually perform the vector calculations:

$$\overline{X_T} = \frac{\overline{X_L} \times \overline{X_C}}{\overline{X_L} + \overline{X_C}} \tag{19.9}$$

$$\overline{X_T} = \frac{24\angle 90° \times 10\angle -90°}{24\angle 90° + 10\angle -90°} \tag{19.10}$$

Performing this vector calculation on a scientific calculator produces the following result:

$$\overline{X_T} = 17.14\angle -90° \tag{19.11}$$

Notice that the magnitude is the same as calculated using method 1; further, the $-90°$ angle agrees with the previous decision that the total reactance is capacitive.

19.3 Current

The total current flow can be calculated using Ohm's law:

$$I = \frac{E}{Z}$$

$$I = \frac{220 \text{ V}}{17.14 \text{ }\Omega} = 12.84 \text{ A}$$

Since this is a parallel circuit, the voltage across each element is the same; consequently, the current through each element is calculated using Ohm's law:

FIGURE 19–10 Current flow in Figure 19–7

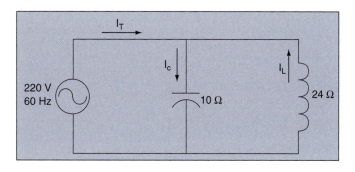

$$I_L = \frac{220 \text{ V}}{24 \text{ }\Omega} = 9.17 \text{ A} \qquad (19.12)$$

$$I_C = \frac{220 \text{ V}}{10 \text{ }\Omega} = 22 \text{ A} \qquad (19.13)$$

The total current can now be checked using Figure 19–10. From Kirchhoff's current law, you know that the sum of all the currents flowing into a node must be equal to zero. In Figure 19–10, this can be written mathematically as

$$I_T = I_C - I_L \qquad (19.14)$$

and

$$I_T = 22 \text{ A} - 9.17 \text{ A} = 12.83 \text{ A} \qquad (19.15)$$

Notice that the current has the same sign as I_C, meaning that the capacitive current is the larger of the two.

RESONANT CIRCUITS

19.4 Series Resonant Circuits

When an inductor and capacitor are connected in series, there is a frequency at which the inductive reactance and capacitive reactance will become equal. This is because, as frequency increases, inductive reactance increases and capacitive reactance decreases. The point at which two reactances become equal is called *resonance*. Resonant circuits can be used to provide increases of current and voltage at the **resonant frequency**. The formula used to determine the resonant frequency when the values of L and C are known is as follows:

$$f_R = \frac{1}{2\pi\sqrt{LC}}$$

Where:

f_R = frequency at resonance
L = inductance in henrys
C = capacitance in farads.

EXAMPLE 2

Find the resonant frequency for a series circuit with an inductance of .0154 H and a capacitance of 1.61 μF.

Solution:

$$f_R = \frac{1}{2\pi\sqrt{LC}}$$

$$f_R = \frac{1}{2 \times 3.14 \times \sqrt{0.0154 \times 0.00000161}}$$

$$f_R = \frac{1}{6.28(0.00016)}$$

$$f_R = 1{,}011 \text{ Hz}$$

This circuit will reach resonance at 1,011 Hz when both the inductor and the capacitor produce equal reactances. At this frequency the two reactances are going in opposite directions, and since they are equal, they cancel each other out.

When a circuit is not at resonance, current flow is limited by the combination of inductive reactance and capacitive reactance. At a lower frequency, the inductive reactance decreases; however, since the capacitive reactance increases, the total impedance increases.

If the frequency is increased above the resonant frequency, the inductive reactance increases, the capacitive reactance decreases, and the total impedance increases. In either case, there is more impedance than when both reactances cancel each other.

The other issue is the effect of resonance on current. Look at Figure 19–11. When the circuit reaches resonance (the frequency at which both *L* and *C* reactance are equal), the current will suddenly increase because the only current-limiting factor will be wire resistance.

FIGURE 19–11 Current increases at resonant frequency.

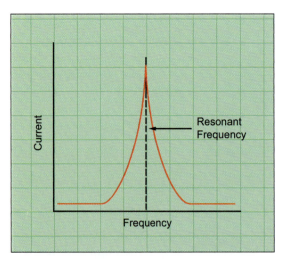

19.5 Parallel Resonant Circuits

Remember that when inductive reactance and capacitive reactance become equal, the circuit is said to be resonant. In a parallel circuit, inductive current and capacitive current cancel each other out because they are the parameters that are opposing each other; that is, they are 180° out of phase with each other. When a parallel circuit reaches resonance, the total circuit current should be zero, and the total circuit impedance should become infinite. This produces a minimum line current at the point of resonance. LC parallel circuits are often referred to as *tank circuits*.

Remember that the voltage drop across all parts of a parallel circuit is the same.

EXAMPLE 3

Given the circuit in Figure 19–12 and the values E = 200 VAC, X_C = 50 ohms, and X_L = 40 ohms, what is I_L, I_C, and I_T?

Solution:

$$I_L = \frac{E}{X_L}$$

$$I_L = \frac{200}{40}$$

$$I_L = 5 \text{ A}$$

$$I_C = \frac{E}{X_C}$$

$$I_C = \frac{200}{50}$$

$$I_C = 4 \text{ A}$$

$$I_T = I_L - I_C$$

$$I_T = 5 - 4$$

$$I_T = 1 \text{ A}$$

Given the parallel circuit shown in Figure 19–12 and the values 6 volts, 350 Hz, capacitor = .57 μF, and inductance = 105 mH, find X_L, X_C, I_L, I_C, and I_T.

Solution:

$$X_L = 2\pi fL$$

$$X_L = 2 \times 3.14 \times 350 \times 0.105$$

$$X_L = 231 \; \Omega$$

$$X_C = \frac{1}{2\pi fC}$$

$$X_C = \frac{1}{2 \times 3.14 \times 350 \times .00000057}$$

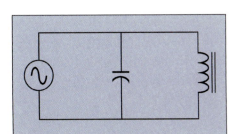

FIGURE 19–12   Parallel LC circuit for Example 3.

$$X_C = 798 \ \Omega$$

$$I_L = \frac{E}{X_L}$$

$$I_L = \frac{6}{231}$$

$$I_L = 0.026 \text{ A, or } 26 \text{ mA}$$

$$I_C = \frac{E}{X_C}$$

$$I_C = \frac{6}{798}$$

$$I_C = .0075 \text{ A, or } 7.5 \text{ mA}$$

$$I_T = I_L - I_C$$

$$I_T = 26 - 7.5$$

$$I_T = 18.5 \text{ mA}$$

Remember that, for a parallel circuit, as frequency decreases, X_C will increase, and X_L will decrease.

■ SUMMARY

The voltage drop across the inductor in a series LC circuit will lead the current by 90°, and the voltage drop across the capacitor in a series LC circuit will lag the current by 90°. In an series LC circuit, inductive and capacitive values are 180° out of phase with each other and results in a voltage that is less than the larger of the two components.

The current flow through the inductor in a parallel LC circuit will lag the applied voltage by 90°, and the current flow through the capacitor in a parallel LC circuit will lead the applied voltage by 90°. Adding them will result in a total current that is smaller than the larger of the two components.

LC resonant circuits increase the current and voltage drop across the individual elements at the resonant frequency. Resonance occurs at the frequency where inductive reactance and capacitive reactance are equal to each other.

■ REVIEW QUESTIONS

1. Using the current as a reference, discuss the voltage drops and phase angles in a series LC circuit.

2. Using the voltage as a reference, discuss the current flows and phase angles in a parallel LC circuit.

3. Draw a series RLC circuit and the approximate vector diagram for the voltage drops in the circuit.

4. Draw a parallel RLC circuit and the approximate vector diagram for the current flow in the circuit.

5. Discuss the formula for calculating resonant frequency for an LC circuit.

6. What happens to capacitive reactance and inductive reactance when the circuit frequency goes up? Goes down?

PRACTICE PROBLEMS

Problems 1 to 3 refer to the following figure:

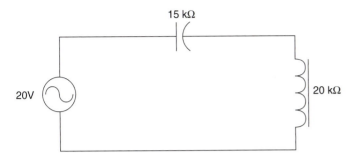

1. Find I.

2. If E_T goes up to 36 volts, what is the new current?

3. If the frequency doubles at 20 volts, what is the new current?

The remaining problems refer to the following figure:

4. Find I_L.

5. Find I_C.

6. If the frequency doubles, find I_L.

7. Find I_C.

8. Find I_T.

9. Using the currents calculated in problems 4 and 5, draw a current vector diagram of the circuit.

chapter 20

Resistive, Inductive, and Capacitive (RLC) Circuits in Series

■ OVERVIEW

Y ou now know that inductance causes current in the circuit to lag the applied voltage and that capacitance causes the current to lead the applied voltage. The voltage across an inductor leads the current by 90°, and the voltage across the capacitor lags the current by 90°.

Remember that current in a series circuit is the same at all points and therefore can be used as a reference to analyze the circuit. In a parallel circuit, voltage will be the same across each component, or leg, and is used as the reference. As stated before, using current for the reference is for learning purposes. Most professional vector diagrams use voltage as the reference.

From these principles, you conclude that in a series inductive capacitive (LC) circuit, X_L and X_C are 180° out of phase with each other. Therefore, the value of one will subtract from the other, and the remainder will determine if the circuit is either inductive or capacitive. In a parallel LC circuit, I_L and I_C are 180° out of phase, with the larger one determining whether the circuit is inductive or capacitive.

This chapter introduces resistance into the formula. With resistance present, the total impedance will have a resistive as well as a capacitive or inductive component.

■ OBJECTIVES

After completing this chapter, you should be able to:

1. Describe the behavior of an RLC circuit when parameters are changed.
2. Mathematically analyze a series RLC circuit for given parameters.

■ RLC SERIES CIRCUITS

When an AC circuit contains the elements of resistance, capacitance, and inductance connected in series, the current is the same through all parts, but the voltage drops across the components are out of phase with each other. This is a characteristic of a series circuit.

20.1 Voltage Drop

The voltage drop across the resistor will be in phase with the current. The voltage drop across the inductor will lead the current by 90°, and the voltage drop across the capacitor will lag the current by 90°. The three sine wave diagrams in Figure 20–1 show this.

The ratio of total reactance to resistance will determine how much the applied voltage will lag or lead the circuit current. If the circuit is more inductive than capacitive, the current will lag the applied voltage, and the power factor will be a lagging power factor. If the circuit is more capacitive than inductive, the current will lead the voltage, and the power factor will be a leading power factor.

Look at Figure 20–2. The sine waves, shown at the top, represent total current and the voltage across each element. Notice that E_R is in phase with the current, while E_C and E_L are 90° in the lag and the lead, respectively. This means the the inductive voltage and the capacitive voltage are 180° out of phase with each other.

The vectors representing those component values are also 180° out of phase, as seen in the vector diagram to the lower right of the figure. These relationships are critical to understanding how the series RLC circuit components interact and perform.

Since inductive reactance and capacitive reactance are 180° out of phase with each other, they subtract from each other in an AC circuit. This subtraction can mean a total reactance that is less than the greater

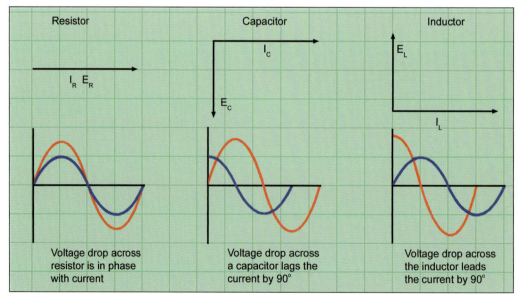

FIGURE 20–1   Voltage and current relationships in RLC series circuits.

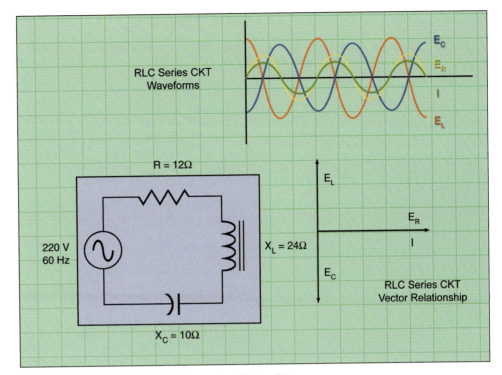

FIGURE 20–2   RLC series circuit relationships.

of the two and perhaps even zero. This means that the current may be very high, depending on the value of R. Also, when Ohm's law is used on the circuit values, the voltage drops developed across these components can be higher than the applied voltage.

20.2 Total Impedance

The impedance of the circuit is the vector sum of the resistance, inductive reactance, and capacitive reactance. Since the inductive reactance and capacitive reactance are 180° out of phase with each other, vector addition must be used to find their sum. This process will result in the smaller of the two being subtracted from the larger. The smaller value is eliminated, and the larger value is reduced by the value of the smaller.

After the total reactance is determined, the total impedance must be calculated using the Pythagorean theorem:

$$Z = \sqrt{R^2 + (X_L - X_C)^2}$$

The total impedance for the circuit of Figure 20–2 can be found as follows:

$$Z = \sqrt{R^2 + (X_L - X_C)^2}$$

$$Z = \sqrt{12^2 + (24 - 10)^2}$$

$$Z = \sqrt{144 + 196}$$

$$Z = 18.4 \ \Omega$$

20.3 Current

The total current flow through the circuit is calculated using Ohm's law:

$$I = \frac{E}{Z}$$

$$I = \frac{220}{18.4}$$

$$I = 12.0 \text{ A}$$

The current flow is the same throughout the series circuit, so 12.0 A will flow through all the components.

■ RESISTIVE CALCULATIONS

20.4 Resistive Voltage Drop

The voltage drop across the resistor can be calculated using the following formula:

$$E_R = I \times R$$

$$E_R = 12.0 \times 12$$

$$E_R = 144 \text{ V}$$

20.5 Watts Dissipation

The true power of the circuit is calculated using any of the pure resistive values. Remember that true power exists only when current and voltage are in phase with each other, and that happens only in the pure resistive parts of the circuit. For this example, use the following formula:

$$P = E_R \times I$$

$$P = 144 \times 12.0$$

$$P = 1,728.0 \text{ W}$$

■ INDUCTIVE CIRCUIT CHARACTERISTICS

20.6 Inductance

The inductance of the circuit can be calculated using the following formula:

$$X_L = 2\pi fL$$

$$L = \frac{X_L}{2\pi f}$$

$$L = \frac{24}{2 \times 3.14 \times 60}$$

$$L = \frac{24}{376.8}$$

$$L = 0.0637 \text{ H}$$

20.7 Voltage Drop Across the Inductor

The amount of voltage drop across the inductor can be calculated using the following formula:

$$E_L = I \times X_L$$
$$E_L = 12.0 \times 24$$
$$E_L = 288 \text{ V}$$

■ CAPACITIVE CIRCUIT CHARACTERISTICS

20.8 Capacitance

The amount of capacitance in the circuit can be calculated using the following formula:

$$C = \frac{1}{2\pi f X_C}$$

$$C = \frac{1}{2 \times 3.14 \times 60 \times 10}$$

$$C = 0.0002654 \text{ F, or } 265.4 \text{ }\mu\text{F}$$

20.9 Voltage Drop Across the Capacitor

The voltage drop across the capacitor is calculated using the following formula:

$$E_C = I \times X_C$$
$$E_C = 12.0 \times 10$$
$$E_C = 120 \text{ V}$$

■ RLC CIRCUIT POWER CHARACTERISTICS

20.10 Apparent Power

Volt-amps can be calculated by multiplying the applied voltage times the circuit current:

$$VA = E_T \times I$$
$$VA = 220 \times 12.0$$
$$VA = 2,640$$

20.11 Power Factor

The power factor is calculated by dividing the true power of the circuit by the apparent power. The answer is multiplied by 100 to turn the decimal into a percentage:

$$PF = \frac{W}{VA} \times 100$$

$$PF = \frac{1,728}{2,640} \times 100$$

$$PF = 0.65 \times 100$$

$$PF = 65\%$$

20.12 Angle Theta

The power factor is the cosine of angle theta:

$$\cos \angle \theta = .65$$
$$\angle \theta = 49.5°$$

■ SUMMARY

The voltage drop across the resistor in a series RLC circuit is in phase with the current. The voltage drop across the inductor in a series RLC circuit will lead the current by 90°, and the voltage drop across the capacitor in a series RLC circuit will lag the current by 90°. The current is the same at all points in a series circuit.

Vector addition can be used in a series RLC circuit to find values of total voltage, impedance, and apparent power. In an RLC circuit, inductive and capacitive values are 180° out of phase with each other. The smaller value is subtracted from the larger value, resulting in a reduced larger value.

■ REVIEW QUESTIONS

1. What are voltage relationships in a series RLC circuit?
2. What are the current relationships in a series RLC circuit?
3. Discuss how you would calculate the following:
   a. Total impedance
   b. Total current flow
   c. Resistive voltage
   d. Inductive voltage
   e. Capacitive voltage
   f. Power factor

■ PRACTICE PROBLEMS

Problems 1 to 5 refer to the following figure:

Problems 6 to 10 refer to the following figure:

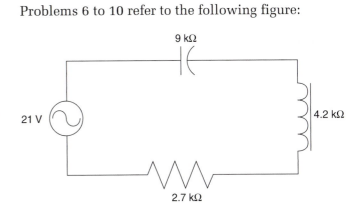

1. Find the impedance.
2. Find E_T.
3. Find E_R, E_C, and E_L.
4. What is the true power of the circuit?
5. What is the phase angle and PF?

6. Find Z.
7. Find I.
8. Find E_R, E_L, and E_C.
9. What power is used in this circuit? Prove it.
10. Find E_T and draw the vector diagram.

chapter 21

Resistive, Inductive, and Capacitive (RLC) Circuits in Parallel

■ OUTLINE

■ OVERVIEW

Most of the circuits you will work with in your professional career will contain resistance, inductance, and capacitance. Most AC loads tend to be inductive because of the heavy presence of inductive loads like motors and fluorescent light fixtures. Further, most of the circuits you will work with will be a combination of series and parallel. The load circuits themselves are usually primarily RL in series and are connected in parallel between the supply wires.

This chapter continues the work you did in the previous chapter. Remember that in a parallel circuit, the voltage is the same across all branches, and the current divides among the branches.

■ OBJECTIVES

After completing this chapter, you should be able to:

1. Describe the characteristics of a parallel RLC circuit when parameters are changed.
2. Calculate for unknown parameters when sufficient information is given.

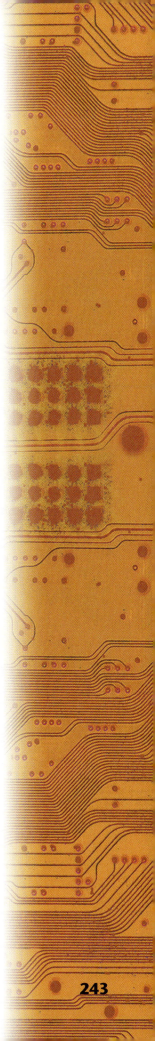

■ PARALLEL RLC CIRCUITS

When an AC circuit has the elements of resistance, inductance, and capacitance connected in parallel, the voltage drop across each branch is the same. The magnitude and phase angle of the currents flowing through each branch are out of phase with each other as determined by the resistance or reactance of the branch.

Look at Figure 21–1. The current flowing through the resistive component is in phase with the applied voltage. The current flowing through the inductive branch will lag the applied voltage by 90°, and the current flowing through the capacitive leg will lead the applied voltage by 90°. The phase angle difference between the applied voltage and the total current is determined by the ratio of resistance, inductance, and capacitance connected in parallel.

In a parallel RLC circuit, if the inductance is greater, so is the X_L, causing more of the available circuit current to flow through the capacitive branch. The circuit would be more capacitive, and the current would lead the voltage.

■ CALCULATING PARALLEL RLC CIRCUIT VALUES

Look at Figure 21–2. The following problems are solved using the circuit values shown on the drawing.

21.1 Impedance

Remember from previous chapters that the impedance of the parallel circuit is the reciprocal of the sum of the reciprocals of the branches. Vector addition must be used because the values are out of phase:

$$Z = \frac{X_\text{T} \times R}{\sqrt{X_\text{T}^2 + R^2}}$$

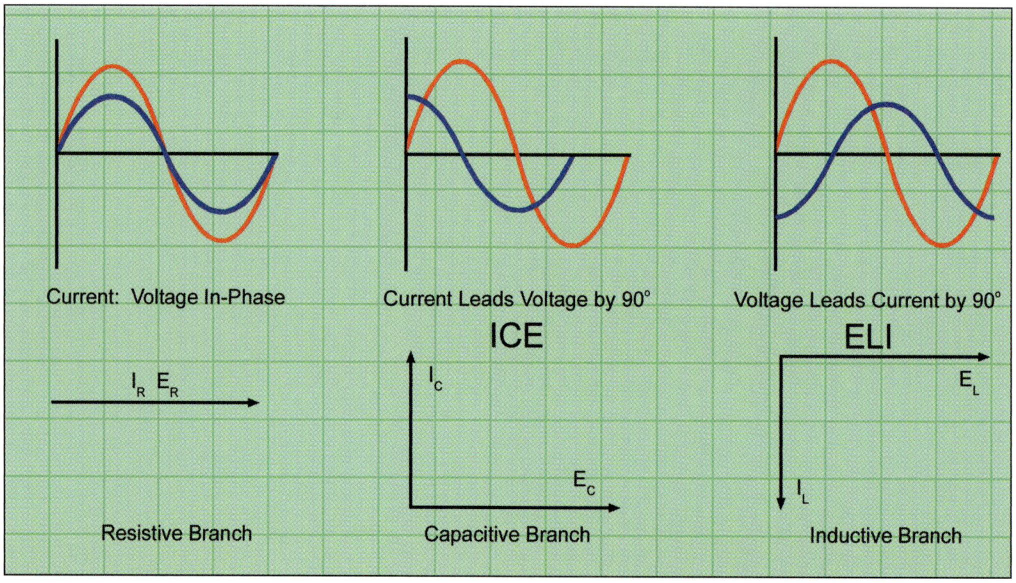

FIGURE 21–1   Current and voltage relationships of parallel RLC circuits.

FIGURE 21–2 Parallel RLC circuit.

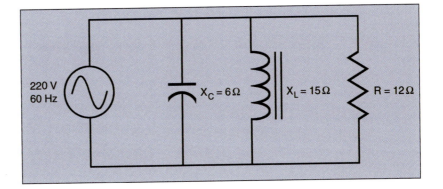

where

$$X_T = \frac{X_L \times X_C}{X_L + X_C}$$

Important: When solving for X_T, the X_L is a positive number, and X_C is a negative number. It is vitally important that this concept be followed as shown here:

$$X_T = \frac{15 \times (-6)}{15 + (-6)}$$

$$X_T = \frac{-90}{9}$$

$$X_T = -10$$

The minus value shows that the net result is a capacitive reactance. Because the next part of the calculations use the Pythagorean theorem, the negative sign will not be used:

$$Z = \frac{X_T \times R}{\sqrt{X_T^2 + R^2}}$$

$$Z = \frac{10 \times 12}{\sqrt{(10)^2 + (12)^2}}$$

$$Z = \frac{120}{\sqrt{100 + 144}}$$

$$Z = \frac{120}{\sqrt{244}}$$

$$Z = \frac{120}{15.62}$$

$$Z = 7.68 \ \Omega$$

21.2 Resistive Current

The next unknown value to calculate is the current flow through the resistor. This can be calculated using the formula

$$I_R = \frac{E}{R}$$

$$I_R = \frac{220}{12}$$

$$I_R = 18.3 \ \text{A}$$

21.3 True Power

Remember that true power can be found only in the resistive leg of the circuit. The true power or watts can be calculated using the formula

$$P = E \times I_R$$
$$P = 220 \times 18.3$$
$$P = 4{,}026 \text{ W}$$

21.4 Inductive Current

The amount of current flow through the inductor is calculated using the formula

$$I_L = \frac{E}{X_L}$$
$$I_L = \frac{220}{15}$$
$$I_L = 14.67 \text{ A}$$

21.5 Inductance

The amount of inductance in the circuit can be calculated using the formula

$$X_L = 2\pi f L$$
$$L = \frac{X_L}{2\pi f}$$
$$L = \frac{15}{2 \times 3.14 \times 60}$$
$$L = \frac{15}{377}$$
$$L = 0.397 \text{ H}$$

21.6 Capacitive Current

The current flow through the capacitor can be found using the formula

$$I_C = \frac{E}{X_C}$$
$$I_C = \frac{220}{6}$$
$$I_C = 36.67 \text{ A}$$

21.7 Capacitance

The value of circuit capacitance can be calculated using the formula

$$C = \frac{1}{2\pi f X_C}$$
$$C = \frac{1}{2 \times 3.14 \times 60 \times 6}$$
$$C = \frac{1}{2{,}260.8}$$
$$C = .00044 \text{ A}$$

■ TOTAL CIRCUIT CALCULATIONS

21.8 Total Current

The value for total current flow in the circuit can be calculated using vector addition of the current flow through each branch of the circuit. Look at Figure 21–3. As shown in Figure 21–3a, the inductive current is 180° out of phase with the capacitive current. These two currents tend to cancel each other out the same way opposing reactances did in the series circuit. You will subtract the smaller from the larger. The total circuit current is the hypotenuse of the resultant right triangle.

The following steps show how the total current in Figure 21–3b was calculated:

$$I_T^2 = I_R^2 + (I_C - I_L)^2$$
$$I_T^2 = (18.3)^2 + (36.67 - 14.67)^2$$
$$I_T^2 = 334.89 + (22)^2$$
$$I_T^2 = 344.89 + 484$$
$$I_T^2 = 818.89$$
$$I_T = \sqrt{818.89}$$
$$I_T = 28.62 \text{ A}$$

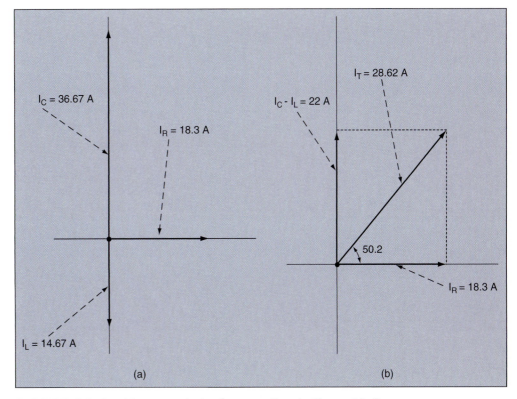

FIGURE 21–3   Vector analysis of current flow in Figure 21–2.

A second way to calculate total current is to use the impedance value calculated earlier:

$$I_T = \frac{E}{Z}$$

$$I_T = \frac{220}{7.68}$$

$$I_T = 28.65 \text{ A}$$

Note the slight difference in the current values due to rounding.

21.9 Apparent Power

Now that the total circuit current has been calculated, the apparent power or volt-amps (VA) can be calculated using the total current in the following formula:

$$VA = E \times I_T$$

$$VA = 220 \times 28.62$$

$$VA = 6{,}296$$

21.10 Power Factor

Since the volt-amps are known, the power factor (PF) can be determined using the following formula:

$$PF = \frac{\text{true power}}{\text{apparent power}}$$

$$PF = \frac{W}{VA} \times 100$$

$$PF = \frac{4{,}026}{6{,}296} \times 100$$

$$PF = .64 \times 100$$

$$PF = 64\%$$

21.11 Angle Theta

Remember that the power factor is the cosine of angle theta. Therefore, angle theta is

$$\cos \theta = 0.64$$

$$\theta = 50.2$$

■ PARALLEL RESONANT CIRCUITS

You know that in a parallel circuit, the inductive current and capacitive current are 180° out of phase and subtract from each other. This produces a minimum line current at the point of resonance. This minimum line current produces maximum impedance. In theory, when a parallel LC circuit reaches resonance, the total circuit current should

FIGURE 21–4 The effects of parallel resonance on LC circuit current.

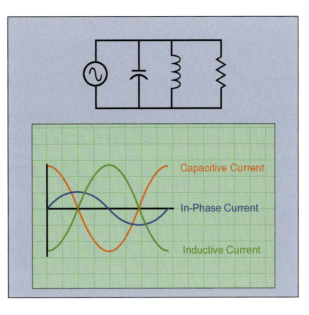

FIGURE 21–5 Tank circuit.

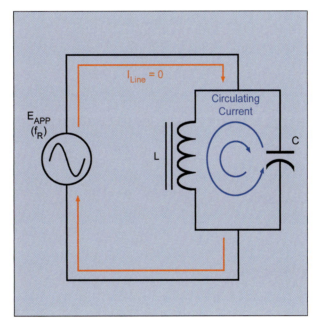

reach zero. Figure 21–4 shows this relationship. At resonance frequency (f_R), the capacitive reactive current and inductive reactance current are equal and cancel.

Figure 21–5 is an LC resonant circuit called a "tank" circuit. Note that the current in the inductor or capacitor branch of the circuit is very high at resonance when compared to the total line current. The individual branch currents cancel the main line current because each is 90° out of phase with the generated line current and opposite each other. As the AC-generated voltage reverses polarity, the stored energy in the inductor and capacitor discharge back and forth across each other. This cycling effect generates a sine wave at the resonant frequency of the LC parallel branches. This sine wave generation is called the *flywheel effect* of an LC tank circuit.

As you learned earlier, an increase in frequency decreases capacitive reactance, increases total current, and decreases impedance. An increase in frequency also increases the power factor.

■ DETERMINING IMPEDANCE USING THE ASSUMED VOLTAGE METHOD

Given the parallel circuit shown in Figure 21–4 and assuming the following values: $X_L = 24\ \Omega$, $X_C = 8\ \Omega$, and $R = 6\ \Omega$, here is how the assumed voltage method works:

1. When possible, assume a voltage that will result in currents for all three legs that are whole numbers. "Assuming" means that we will simply use a number that easily divides by the resistive, inductive, and capacitive values.

2. A voltage of 24 V is assumed for this circuit.

3. Using Ohm's law for AC circuits, the current through each leg will be determined:

$$I_L = \frac{E}{X_L} \qquad I_C = \frac{E}{X_C} \qquad I_R = \frac{E}{R}$$

$$I_L = \frac{24}{24} \qquad I_C = \frac{24}{8} \qquad I_R = \frac{24}{6}$$

$$I_L = 1\text{ A} \qquad I_C = 3\text{ A} \qquad I_R = 4\text{ A}$$

4. The total current for the circuit can now be determined for the assumed voltage of 24 V:

$$I_T = \sqrt{I_R^2 + (I_C - I_L)^2}$$

$$I_T = \sqrt{(4)^2 + (3 - 1)^2}$$

$$I_T = \sqrt{(4)^2 + (2)^2}$$

$$I_T = \sqrt{16 + 4}$$

$$I_T = \sqrt{20}$$

$$I_T = 4.472\text{ A}$$

5. With the 24-V power supply applied, the circuit current is 4.472 A. The impedance of the circuit can now be determined by using Ohm's law:

$$Z = \frac{E}{I_T}$$

$$Z = \frac{24}{4.472}$$

$$Z = 5.367\ \Omega$$

This method is sometimes easier than the reciprocal method for determining the impedance of a parallel circuit. To see how assuming a different voltage supply affects the circuit, substitute 12 V for the assumed voltage and then assume 48 V. What are the outcomes?

$$\text{at 12 V}, Z = 5.367 \ \Omega$$
$$\text{at 48 V}, Z = 5.367 \ \Omega$$

■ SUMMARY

The voltage across all legs of a parallel RLC circuit is the same. The current flow in the resistive leg will be in phase with the voltage. The current flow through the inductive leg will lag the voltage by 90° (ELI), and the current flow through the capacitive branch will lead the voltage by 90° (ICE).

The impedance (Z_T) is equal to the applied voltage divided by the total circuit current (I_T). Z_T is max-

imum at resonance frequency (f_R) because the I_T is minimum. A resonant LC tank circuit is very useful for generating a sine wave at the tank's resonant frequency. This is called the flywheel effect. The tank circuit is also useful in generating a very high load of output voltage at resonance.

■ REVIEW QUESTIONS

1. In a parallel RLC circuit, what is the relationship between the following?
   a. Applied voltage and branch voltage
   b. Total current and individual branch current
   c. Voltage and current of each individual branch
2. In Figure 21–3b, why is the Pythagorean theorem used to calculate the total current?

3. Power factor is the ratio of ____ to ____. Explain.
4. What determines the phase angle between the total current and the applied voltage in a parallel RLC circuit?
5. A tank circuit operates on the so-called flywheel effect. Explain.

■ PRACTICE PROBLEMS

All the practice problems refer to the following figure. Find the total reactance in the following figure.

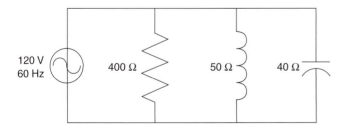

1. Find Z.
2. Find the current in each branch.

3. Find I_T.
4. Using E_T and I_T, find Z. Does this answer agree with question 2?
5. What is the true power in this circuit?
6. What is the apparent power (VA)?
7. What is the power factor?
8. What is the phase angle?
9. Find the Z using the assumed voltage method. Does it agree with question 2?
10. Draw the vector diagram.

chapter 22

Comparing Resistive, Inductive, and Capacitive (RLC) Circuits in Series and Parallel

■ **OUTLINE**

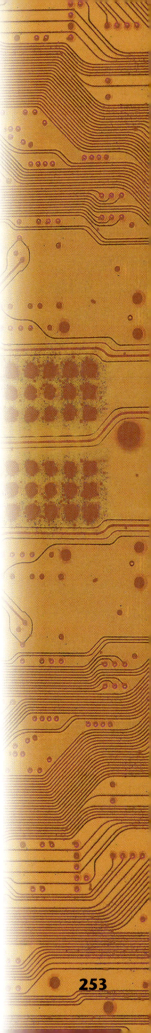

■ OVERVIEW

By now, you should understand that inductance causes current in the circuit to lag the applied voltage and that capacitance causes the current to lead the applied voltage. The voltage across an inductor leads the current by 90°, and the voltage across the capacitor lags the current by 90°.

Remember that current in a series circuit is the same at all points and therefore can be used as a reference to analyze the circuit. In a parallel circuit, voltage will be the same across each component, or leg, and is used as the reference.

From this, you can conclude that in a series inductive capacitive (LC) circuit, X_L and X_C are 180° out of phase with each other. Therefore, the value of one will subtract from the other, and the remainder will determine whether the circuit is either inductive or capacitive. In a parallel LC circuit, I_L and I_C are 180° out of phase, with the larger one determining whether the circuit is inductive or capacitive.

■ OBJECTIVES

After completing this chapter, you should be able to:

1. Explain completely the changes in series and parallel RLC circuits when a circuit parameter is changed.
2. Use vectors and graphs to depict the parameters in these circuits.

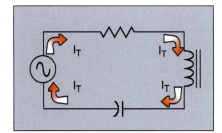

FIGURE 22–1 Series RLC circuit.

■ SERIES AND PARALLEL RLC CIRCUITS COMPARED

22.1 Series RLC Circuits

When an AC circuit contains the elements of resistance, inductance, and capacitance connected in series, the current is the same through all parts (see Figure 22–1), but the voltage drops across the components are out of phase with each other. This is a characteristic of a series circuit. The voltage drop across the resistor is in phase with the current; consequently, the resistor dissipates true power. The voltage drop across the inductor will lead the current by 90°, and the voltage drop across the capacitor will lag the current by 90°. The three sine wave diagrams in Figure 22–2 show the three waveforms.

The relative values of resistance, inductance, and capacitance will determine how much the current will lag or lead the applied voltage. If the circuit is more inductive than capacitive, the current will lag the applied voltage, and the power factor will be a lagging power factor. If the circuit is more capacitive than inductive, the current will lead the voltage, and the power factor will be a leading power factor.

Since inductive reactance and capacitive reactance are 180° out of phase with each other, they subtract from each other in an AC circuit. This subtraction can allow the impedance to become less than the largest or both of the reactances. If the two values are equal, they will completely cancel, and a high amount of current will result. Also, when applying Ohm's law for circuit values, the voltage drops developed across the inductance and capacitance can be higher than the applied voltage.

22.2 Parallel RLC Circuits

When an AC circuit has the elements of resistance, inductance, and capacitance connected in parallel, the rules associated with a parallel cir-

FIGURE 22–2 Series RLC circuit using current as the reference.

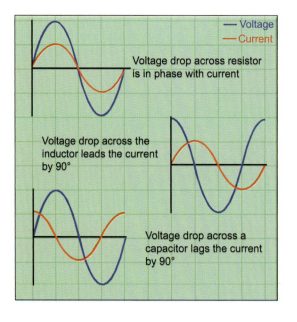

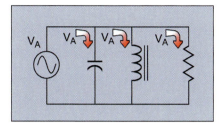

FIGURE 22–3   Parallel RLC circuit.

cuit must be applied. Remember that the voltage drop across each leg, or component, is the same (see Figure 22–3), and the currents flowing through each branch, or leg, will be out of phase with each other.

The current flowing through the resistive component will be in phase with the applied voltage. The current flowing through the inductive part will lag the applied voltage by 90°, and the current flowing through the capacitive leg will lead the applied voltage by 90°. The three sine waves in Figure 22–4 show this relationship.

The phase angle difference between the applied voltage and the total current is determined by the relative sizes of the resistance, inductance, and capacitance connected in parallel. In a series RLC circuit, if the inductance is greater than the capacitance, the current will lag the voltage, and the power factor will be lagging. If the capacitance is greater than the inductance, the current will lead the voltage, and the power factor will be leading.

The opposite is true for parallel circuits. A larger X_L means more current flow in the capacitive branch; therefore, the current leads the applied voltage (see Table 22–1).

In X_L, the frequency value is multiplied, so the increase in reactance is directly proportional to the frequency. In X_C, the frequency value is in the denominator, or "divided by," so an increase in the frequency decreases reactance, and its effect is called *inversely proportional.*

■ SERIES PROBLEM SOLVING

All the following sections refer to the circuit and values shown in Figure 22–5.

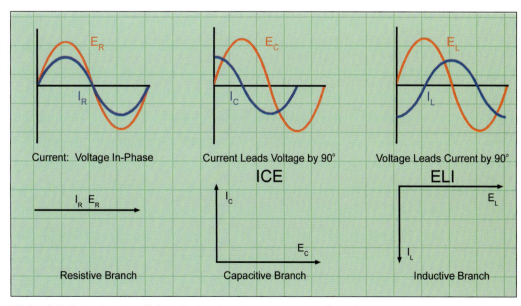

FIGURE 22–4   Parallel RLC circuit using current as the reference.

Table 22–1   Comparison Summary Between Series RLC and Parallel RLC Circuits

RLC Characteristics	Series RLC Circuit	Parallel RLC Circuit
Source reference value	Current	Voltage
X_C greater than (>) X_L	Circuit considered capacitive, and line 1 will lead line V	Circuit considered inductive, and line 1 will lag line V
Frequency increase[a]	X_L will increase	X_L will increase
	X_C will decrease	X_C will decrease
Impedance (Z in Ω)	Z is minimum when $X_C = X_L$	Z is maximum when $X_C = X_L$
Voltage and current	Voltage drop on individual components can be greater than source voltage	Current through any single branch can be greater than source current

[a]Remember the formulas for X_L and X_C: $X_L = 2\pi fL$ and $X_C = \dfrac{1}{2\pi fC}$.

FIGURE 22–5   Series RLC circuit.

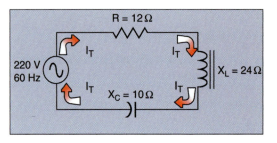

22.3 Impedance

The total impedance is determined using the Pythagorean theorem for vector addition of the magnitudes:

$$Z = \sqrt{R^2 + (X_L - X_C)^2}$$
$$Z = \sqrt{12^2 + (24 - 10)^2}$$
$$Z = \sqrt{144 + 196}$$
$$Z = \sqrt{340}$$
$$Z = 18.4\ \Omega$$

22.4 Current

The total current flow through the circuit is calculated using the variation of Ohm's law formula:

$$I = \frac{E}{Z}$$
$$I = \frac{220}{18.4}$$
$$I = 12.0\ A$$

The current flow is the same throughout the series circuit, so 12.0 A will be the current through all the components.

22.5 Resistive Voltage Drop

The voltage drop across the resistor is calculated using the formula

$$E_R = I \times R$$
$$E_R = 12.0 \times 12$$
$$E_R = 144 \text{ V}$$

22.6 Watts

The true power of the circuit is calculated using any of the values associated with the resistor. Remember that true power exists only when current and voltage are in phase with each other; this happens in the pure resistive parts of the circuit. For this example, use the following formula:

$$P = E_R \times I$$
$$P = 144 \times 12.0$$
$$P = 1,728.0 \text{ W}$$

22.7 Inductance

The inductance of the circuit is calculated using the following formula:

$$X_L = 2\pi fL$$
$$L = \frac{X_L}{2\pi f}$$
$$L = \frac{24}{2 \times 3.14 \times 60}$$
$$L = \frac{24}{376.8}$$
$$L = 0.0637 \text{ H}$$

22.8 Voltage Drop Across the Inductor

The amount of voltage drop across the inductor is calculated using the following formula:

$$E_L = I \times X_L$$
$$E_L = 12.0 \times 24$$
$$E_L = 288 \text{ V}$$

22.9 Capacitance

The amount of capacitance in the circuit is calculated using the following formula:

$$C = \frac{1}{2\pi fX_C}$$
$$C = \frac{1}{2 \times 3.14 \times 60 \times 10}$$
$$C = \frac{1}{3,768}$$
$$C = 0.0002654 \text{ F, or } 265.4 \text{ } \mu\text{F}$$

22.10 Voltage Drop Across the Capacitor

The voltage drop across the capacitor is calculated using the following formula:

$$E_C = I \times X_C$$
$$E_C = 12.0 \times 10$$
$$E_C = 120 \text{ V}$$

22.11 Apparent Power

Volt-amps (VA) are calculated by multiplying the applied voltage by the circuit current:

$$VA = E_T \times I$$
$$VA = 220 \times 12.0$$
$$VA = 2,640$$

22.12 Power Factor

The power factor (PF) is calculated by dividing the true power of the circuit by the apparent power. The answer is multiplied by 100 to turn the decimal into a percentage:

$$PF = \frac{\text{true power}}{\text{apparent power}} \times 100$$
$$PF = \frac{W}{VA} \times 100$$
$$PF = \frac{1,728}{2,640} \times 100$$
$$PF = 0.65 \times 100$$
$$PF = 65\%$$

22.13 Angle Theta

The power factor is the cosine of angle theta. Since the PF is .65,

$$\cos \theta = .65$$
$$\theta = 49.5°$$

■ PARALLEL PROBLEM SOLVING

All the following sections refer to the circuit and values shown in Figure 22–6.

FIGURE 22–6 Parallel RLC circuit.

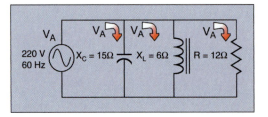

22.14 Impedance

Remember from previous chapters that the impedance of the parallel circuit can be found by using a two-step process. First, the total reactance of the circuit is calculated using the product-over-sum method. Then that value is placed in the impedance problem to solve for the circuit impedance. The following is a step-by-step detail of that process:

$$Z = \frac{X \times R}{\sqrt{X^2 + R^2}}$$

where

$$X = \frac{X_L \times X_C}{X_L + X_C}$$

(remember that X_L is positive and X_C is negative)

$$X = \frac{6 \times (-15)}{6 + (-15)}$$

$$X = \frac{-90}{-9}$$

$$X = 10$$

22.15 Resistive Current

The next unknown value to be determined is the current flow through the resistor. This is calculated using the formula

$$I_R = \frac{E}{R}$$

$$I_R = \frac{220}{12}$$

$$I_R = 18.3 \text{ A}$$

22.16 True Power

Remember that true power can be found only in the resistive leg of the circuit. The true power, or watts, can be calculated using the formula

$$P = E \times I_R$$

$$P = 220 \times 18.3$$

$$P = 4,026 \text{ W}$$

22.17 Inductive Current

The amount of current flow through the inductor is calculated using the formula

$$I_L = \frac{E}{X_L}$$

$$I_L = \frac{220}{6}$$

$$I_L = 36.67 \text{ A}$$

22.18 Inductance

The amount of inductance in the circuit is calculated using the formula

$$L = \frac{X_L}{2\pi f}$$

$$L = \frac{6}{2 \times 3.14 \times 60}$$

$$L = \frac{6}{377}$$

$$L = 0.016 \text{ H}$$

22.19 Capacitive Current

The current flow through the capacitor is found using the formula

$$I_C = \frac{E}{X_C}$$

$$I_C = \frac{220}{15}$$

$$I_C = 14.67 \text{ A}$$

22.20 Capacitance

The value of circuit capacitance is calculated using the formula

$$C = \frac{1}{2\pi f X_C}$$

$$C = \frac{1}{2 \times 3.14 \times 60 \times 15}$$

$$C = 0.00018 \text{ F, or } 180 \text{ }\mu\text{F}$$

22.21 Total Circuit Current

The value for total current flow in the circuit can be calculated using vector addition of the current flow through each branch of the circuit. As shown in Figure 22–7a, the inductive current is 180° out of phase with the capacitive current. These two currents tend to cancel each other out the same way opposing reactances did in the series circuit. You will subtract the smaller from the larger. The total circuit current is the hypotenuse of the remaining, or resultant, right triangle (see Figure 22–7b):

$$I_T = \sqrt{I_R^2 + (I_L - I_C)^2}$$

$$I_T = \sqrt{(18.3)^2 + (36.67 - 14.67)^2}$$

$$I_T = \sqrt{335 + (22)^2}$$

$$I_T = \sqrt{335 + 484}$$

$$I_T = 28.6 \text{ A}$$

The total current could also be calculated using the value of impedance calculated earlier:

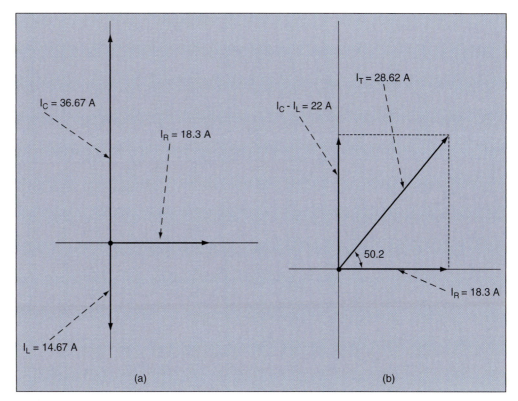

FIGURE 22–7   Vector addition for total current in Figure 22–6.

$$I_T = \frac{E}{Z}$$

$$I_T = \frac{220}{7.68}$$

$$I_T = 28.65 \text{ A}$$

22.22 Apparent Power

Now that the total circuit current has been computed, the apparent power or volt-amps can be calculated using the following formula:

$$\text{VA} = E \times I_T$$
$$\text{VA} = 220 \times 28.65$$
$$\text{VA} = 6,303 \text{ VA}$$

22.23 Power Factor

The power factor is determined using the following formula:

$$\text{PF} = \frac{\text{true power}}{\text{apparent power}} \times 100$$

$$\text{PF} = \frac{W}{\text{VA}} \times 100$$

$$\text{PF} = \frac{4026}{6303} \times 100$$

$$\text{PF} = .64 \times 100$$

$$\text{PF} = 64\%$$

22.24 Angle Theta

Remember that the power factor is the cosine of angle theta. Since the PF is .65,

$$\cos\theta = 0.64$$
$$\theta = 50.2°$$

■ PARALLEL RESONANT CIRCUITS

In a parallel circuit, as shown in Figure 22–8, the inductive current and capacitive current subtract from each other since they are 180° out of phase. This produces a minimum line current at the point of resonance ($X_L = X_C$). In theory, when a parallel LC circuit reaches resonance, the total circuit current should reach zero. This theory assumes that there is no resistor in parallel with the parallel LC.

An increase in frequency decreases capacitive reactance, increases total current, and decreases impedance. An increase in frequency from resonance will decrease the impedance of the LC portion of the RLC circuit. At resonance, the power factor of the tank portion of the RLC circuit is zero; however, the power factor of the entire RLC circuit is 100%. That is, the line current and the applied voltage are in phase. Calculate the total current and impedance for the circuit at resonance, as shown in Figure 22–8.

22.25 Current Calculations

When possible, assume a voltage that will result in currents for all three legs that are whole numbers—in this case, 220 VAC. "Assuming" means using a number that divides easily by the resistive, inductive, and capacitive values. Using Ohm's law for AC circuits, the current through each leg will be determined:

$$I_L = \frac{E}{X_L}$$

$$I_L = \frac{220}{20}$$

$$I_L = 11 \text{ A}$$

$$I_C = \frac{E}{X_C}$$

$$I_C = \frac{220}{20}$$

FIGURE 22–8 Parallel resonant circuit

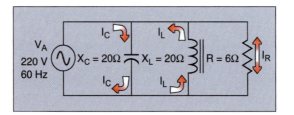

$$I_C = 11 \text{ A}$$

$$I_R = \frac{E}{R}$$

$$I_R = \frac{220}{6}$$

$$I_R = 36.67 \text{ A}$$

The total current for the circuit can now be determined:

$$I_T = \sqrt{I_R^2 + (I_L - I_C)^2}$$

$$I_T = \sqrt{36.67^2 + 0}$$

$$I_T = 36.67 \text{ A}$$

22.26 Finding Impedance

With the 220-V power supply applied, the circuit current is 36.67 A. The impedance of the circuit can now be determined by using Ohm's law:

$$Z = \frac{E}{I_T}$$

$$Z = \frac{220}{36.67}$$

$$Z = 6 \ \Omega$$

This method is usually simpler than the reciprocal method for determining the impedance of a parallel circuit.

■ SUMMARY

This chapter gives you an opportunity to do a point-by-point comparison of the differences between series and parallel RLC circuits. In an RLC circuit, inductive and capacitive values are 180° out of phase with each other. The smaller value is subtracted from the larger value to result in a reduced larger value.

For series RLC circuits;

- The voltage drop across the resistor in a series RLC circuit is in phase with the current.
- The voltage drop across the inductor in a series RLC circuit will lead the current by 90°.
- The voltage drop across the capacitor in a series RLC circuit will lag the current by 90°.
- The current is the same at all points in a series circuit.

- Vector addition can be used in a series RLC circuit to find values of total voltage, impedance, and apparent power.

For parallel RLC circuits:

- The voltage applied to all legs of a parallel RLC circuit is the same.
- The current flow in the resistive leg will be in phase with the voltage.
- The current flow through the inductive leg will lag the voltage by 90°.
- The current flow through the capacitive branch will lead the voltage by 90°.

■ REVIEW QUESTIONS

1. Using Table 22–1, review the differences between parallel and series RLC circuits.

2. At the resonant frequency, what determines the total impedance of a parallel RLC circuit? A series RLC circuit?

3. In a series RLC circuit, how do the voltages on the capacitor and the inductor relate to each other?

4. In a parallel RLC circuit, how do currents through the capacitor and the inductor relate to each other?

5. If the frequency of the applied voltage goes up, how does the total impedance change in a series RLC circuit? A parallel RLC circuit?

■ PRACTICE PROBLEMS

Problems 1 to 6 refer to the following figure:

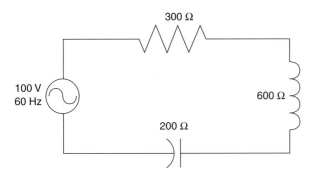

1. Find Z.
2. Find I_T.
3. Find E_R, E_L, and E_C.
4. Find the true power in watts.
5. Find L and C.
6. Find VA, PF, and $\angle\theta$.

Problems 7 to 13 refer to the following figure:

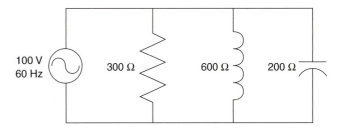

7. Find $Z = X + R\sqrt{R^2 + X^2}$.
8. Find I_R, I_L, I_C, and I_T.
9. Find the true power and the apparent power.
10. Find L and C.
11. Find PF and $\angle\theta$.
12. The series circuit in the first figure is ____. The parallel circuit in the second figure is ____.
13. If frequency is changed so that $X_L = X_C$ in both circuits, then I_T series = ____, and I_T parallel = ____.

chapter **23**

Combination RLC Circuits

■ **OUTLINE**

■ OVERVIEW

Most AC circuits have some combination of resistance, capacitance, and inductance. In fact, most electrical systems have series and parallel combinations of components in the same circuit. Because of the different phase angles involved, vector analysis is the only sure way to analyze a combination circuit of this type.

This chapter teaches you how to use your vector analysis skills in the solution of combination circuits. Remember that all the information you learned previously still applies. Tools such as superposition, Kirchhoff's laws, Thevenin equivalent circuits, and other such methods will work in AC combination circuits. However, the calculations must be made using vector analysis.

■ OBJECTIVES

After completing this chapter, you should be able to:

1. Draw the vectors of the impedance in a complex RLC circuit.
2. Calculate for the unknown values in a complex RLC circuit.

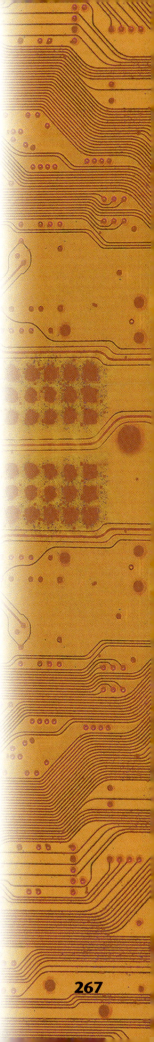

■ COMBINATION RLC CIRCUITS

Using the current for the reference in a series circuit simplifies the analysis. Conversely, the voltage reference is the simplest choice for a parallel circuit. The secret for solving for unknowns in any combination RLC circuit is to reduce the complex circuit to a single equivalent load and source for voltage. This method was taught in the *DC Theory* text and is called the *Thevenin equivalent circuit*. The only differences between DC and AC are that the equivalent resistance in DC must be an equivalent impedance in AC, and the Thevenin voltage may have a phase angle associated with it.

This approach is similar to the approach used in solving for total resistance in series/parallel combinations. The difference is that in an AC circuit containing inductance, capacitance, and resistance, the vector components of the series equivalent circuit will not be in phase with each other. These series vectors must be reduced to their vertical and horizontal components, which can then be added directly together. The sum of the vertical components can be added to the sum of the horizontal components using the Pythagorean theorem.

■ CIRCUIT ANALYSIS: RESISTANCE, INDUCTANCE, AND CAPACITANCE

Consider the procedure to solve for the parameters of combination circuits containing inductance, capacitance, and resistance. Take a look at the circuit in Figure 23–1. Each component or parallel combination is identified as a vector impedance. For example, the 10-Ω series resistor is Z_1, the combination of the 8-Ω resistor and 8-Ω inductor in parallel is Z_2, and so forth around the circuit.

23.1 Parallel Component Vector Addition

In order to add these vector assignments together, the magnitude and direction of the parallel combinations (Z_2 and Z_4) must first be solved. To determine the impedance of each of these vectors, use the assumed voltage method.

For Z_2, assume a voltage of 8 V across the parallel components. Applying the assumed voltage, 1 A will flow through each component. The current through the resistor will be in phase with the voltage, and the current through the inductor will lag the voltage by 90°:

$$I_{Z_2} = \sqrt{I_L^2 + I_R^2}$$
$$I_{Z_2} = \sqrt{1 + 1}$$

Using the Pythagorean theorem, the total current equals the square root of 2, or 1.414 A. The magnitude of Z_2 equals

$$Z_2 = \frac{E_{assumed}}{I_{Z_2}}$$

$$Z_2 = \frac{8\ V}{1.414\ A}$$

$$Z_2 = 5.66\ \Omega$$

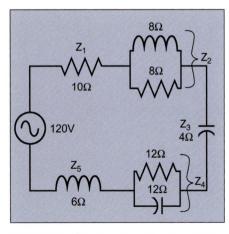

FIGURE 23–1 Combination RLC circuit.

The tangent of θ_{Z_2} is equal to

$$\tan \theta_{Z_2} = \frac{I_L}{I_R}$$

$$\tan \theta_{Z_2} = \frac{1 \text{ A}}{1 \text{ A}}$$

$$\tan \theta_{Z_2} = 1$$

$$\theta_{Z_2} = 45°$$

Therefore, V_2 has an impedance of 5.66 Ω at 45°.

Of course, this same result could be reached quickly with a scientific calculator as follows. For Z_2, you know that

$$\overline{Z_2} = \frac{\overline{X_L} \times \overline{R}}{\overline{X_L} + \overline{R}}$$

where the bar over the Z, X_L, and R means that they are vectors. You also know that

$$\overline{X_L} = 8\angle 90° \text{ and } \overline{R} = 8\angle 0°$$

Substituting,

$$\overline{Z_2} = \frac{8\angle 90° \times 8\angle 0°}{8\angle 90° + 8\angle 0°}$$

Performing this vector calculation with a scientific calculator yields

$$\overline{Z_2} = 5.66\angle 45°$$

Repeat the same process for Z_4, using the assumed voltage method or a scientific calculator to find the Z_4 impedance. Using 12 V as the assumed voltage gives a current through Z_4 of 1.414 A, the same as for Z_2 (you can see how assuming the "good" voltage can make this portion of the problem solving easy).

Using the assumed voltage and newly found current, solve for the impedance of Z_4:

$$Z_4 = \frac{E_{\text{assumed}}}{I_{Z_4}}$$

$$Z_4 = \frac{12 \text{ V}}{1.414 \text{ A}}$$

$$Z_4 = 8.49 \ \Omega$$

Since the value of the current through the resistor and capacitor are both 1 A, the tan θ will be same as the tan θ through the Z_2 group because the current flow through those components was also equal to 1 A. The tangent of that angle is also equal to 1. Since this is a capacitive circuit, the angle will be negative instead of positive; therefore, $\theta = -45°$, and $Z_2 = 8.49\angle -45°$.

If the capacitor and inductor had different values relative to the resistor in the group, the angle theta would calculate to a different magnitude. Because our current through each of the components is equal (the resistance and reactance being equal), the current flow through each component is also equal in our example circuit.

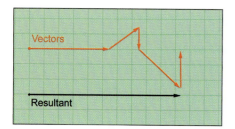

FIGURE 23–2 Vector addition, arrowhead method.

Here is another method. Both Z_2 and Z_4 combinations could be separately solved using the formula

$$Z = \frac{R \times X}{\sqrt{R^2 + X^2}}$$

For example, using Z_2 parameters,

$$Z_2 = \frac{8 \times 8}{\sqrt{8^2 + 8^2}}$$

$$Z_2 = \frac{64}{11.31} = 5.66 \; \Omega$$

23.2 Circuit Total Vector Addition

Figure 23–2 shows the five impedance values of Figure 23–1 added together graphically using the arrowhead-to-end or head-to-tail method. The resultant vector for this circuit is 20 Ω at 0°. Note that Z_3 and Z_5 have no horizontal component since they are a pure capacitance and pure inductance, respectively. Likewise, Z_1 has no vertical component since it is a pure resistance.

This is the solution for this problem. Vectors Z_2 and Z_4 must be reduced to their vertical and horizontal components. Table 23–1 lists the vertical and horizontal components and their calculation. The information in this table is based on the following: $\sin(45°) = .707$, $\sin(-45°) = -.707$, $\cos(45°) = .707$, and $\cos(-45°) = .707$.

The horizontal components of Z_2 and Z_4 can now be combined with Z_1:

$$HC_T = Z_1 + Z_2 + Z_4$$
$$HC_T = 10 \; \Omega + 4 \; \Omega + 6 \; \Omega$$
$$HC_T = 20 \; \Omega$$
$$VC_T = + Z_{2VC} - Z_3 - Z_{4VC} + Z_5$$
$$VC_T = 4 \; \Omega - 4 \; \Omega - 6 \; \Omega + 6 \; \Omega$$
$$VC_T = 0$$

Note that the negative values for Z_3 and Z_4 are due to the capacitive lag of the parallel components.

Table 23–1   **Horizontal and Vertical Vector Components for Figure 23–1**

Component	Z_1 (Ω)	Z_2 (Ω)	Z_3 (Ω)	Z_4 (Ω)	Z_5 (Ω)
Vertical	0	$Z_2 \times \sin(45°) = 4 \; \Omega$	−4	$Z_4 \times \sin(-45°) = -6$	6
Horizontal	10	$Z_2 \times \cos(45°) = 4$	0	$Z_4 \times \cos(-45°) = 6$	0

FIGURE 23-3   Vector addition with values.

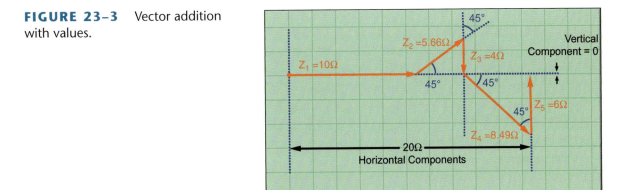

Table 23-2   Voltage Drop Calculations for Figure 23-1

Z_1	Z_2	Z_3	Z_4	Z_5
$E_{Z_1} = I_T \times Z_1$	$E_{Z_2} = I_T \times Z_2$	$E_{Z_3} = I_T \times Z_3$	$E_{Z_4} = I_T \times Z_4$	$E_{Z_5} = I_T \times Z_5$
$E_{Z_1} = 6\,A \times 10\,\Omega$	$E_{Z_2} = 6\,A \times 5.66\,\Omega$	$E_{Z_3} = 6\,A \times 4\,\Omega$	$E_{Z_4} = 6\,A \times 8.49\,\Omega$	$E_{Z_5} = 6\,A \times 6\,\Omega$
$E_{Z_1} = 60\,V$	$E_{Z_2} = 33.96\,V$	$E_{Z_3} = 24\,V$	$E_{Z_4} = 50.94\,V$	$E_{Z_5} = 36\,V$

The mathematical solution confirms the graphic solution (see Figure 23–3). The impedance of the circuit is 20 Ω at 0°. Using this information, the total current for the circuit is determined by Ohm's law:

$$I_T = \frac{E_T}{Z_T}$$

$$I_T = \frac{120\ V}{20\ \Omega}$$

$$I_T = 6\ A$$

Once the circuit current is known, the voltage drop across each impedance vector can be determined using Ohm's law:

$$E_{Z_X} = I_T \times Z_X$$

Table 23–2 summarizes the results.

■ CURRENT SOLUTIONS FOR PARALLEL CIRCUITS

Figure 23–4 shows the vector addition of the voltages. With this information, the currents in the parallel circuits can be determined.

For Z_2, each of the two components has an ohm value of 8 Ω with 33.96 V across them. Each 8-Ω leg of the parallel will have 4.245 A with the current through the resistance leg leading the current through the inductive leg by 90°.

For Z_4, each of the two components has an ohm value of 12 Ω with 50.94 V across them. Each 12-Ω leg will have 4.245 A with the current through the resistive leg lagging the current in the capacitive leg by 90°.

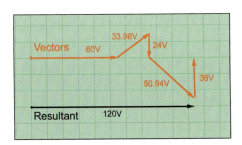

FIGURE 23-4   Current solutions for parallel circuits.

■ SUMMARY

The reference for analyzing any series circuit is the current. The reference for analyzing any parallel circuit is the voltage. In a capacitive circuit, the current leads the voltage. In an inductive circuit, the voltage leads the current.

Capacitive reactance is determined by the frequency and the capacitance. Remember that $X_C = 1/2\pi fC$. Thus, raising the frequency in a capacitive circuit will lower the capacitive reactance (X_C).

Inductive reactance is determined by the frequency and the inductance. Remember that $X_L = 2\pi fL$. Thus, raising the frequency in an inductive circuit will raise the inductive reactance (X_L).

A vector times the sine of theta will give you the vertical component. The horizontal component can be determined by multiplying the vector by the cosine of theta.

Figure 23–5 can be used as a reminder for the various trigonometric values that you may need.

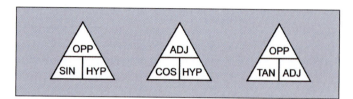

FIGURE 23–5   Triangular relationships.

■ REVIEW QUESTIONS

1. What is the major difference between analyzing combination AC circuits and the simpler series or parallel circuit?

2. In Figure 23–2, vector Z_2 goes up and vector Z_4 goes down. Why?

3. Discuss the use of Kirchhoff's laws, Thevenin equivalent circuits, Norton equivalent circuits, and superposition in RLC combination circuits.

   a. Can they all be used?

   b. What differences would you expect between their use in DC circuits as opposed to AC circuits?

   c. Will arithmetic calculations work, or will the AC circuits require vector calculations?

■ PRACTICE PROBLEMS

Problems 1 to 4 refer to the following figure.

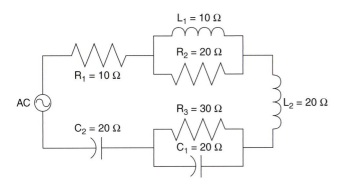

1. Find $Z_{R_2 + L}$ of $L_1 + R_2$ using the assumed voltage method.
2. Find $Z_{R_3 + C}$ of $C_1 + R_3$ by the same method as problem 1.
3. Find Z_{tot}.
4. Diagram the circuit (vectorially).

chapter 24

Series RLC Circuits and Resonance

■ OUTLINE

OVERVIEW

As discussed in earlier chapters, resonance occurs at the frequency where X_L is equal to X_C. This phenomenon of resonance is important to understand for power factor correction and advanced study in this field.

This chapter provides a review of this concept. Remember that current is the same through every part of a series circuit, and we will use it as the reference in this chapter.

OBJECTIVES

After completing this chapter, you should be able to:

1. Explain the operation of a series RLC circuit below, above, and at resonance.
2. Calculate for unknown parameters.
3. Determine Q and the bandwidth of the circuit.

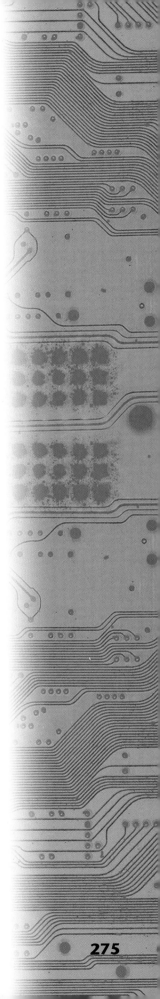

■ RESONANCE

When an inductor and capacitor are connected in series, there is a frequency at which the inductive and capacitive reactances (X_L and the X_C) are equal in magnitude. Since X_L and X_C are 180° out of phase, at resonance they will exactly cancel. This means that at resonance in a series RLC circuit, the total reactance is zero.

24.1 Series Resonant Circuits

When the values of L and C are known, the formula for calculating the value at which resonance occurs is

$$f_R = \frac{1}{2\pi\sqrt{LC}}$$

Where:

f_R = frequency at resonance
L = inductance in henrys
C = capacitance in farads
π = a constant of 3.14.

EXAMPLE 1

Given the circuit shown in Figure 24–1, calculate the resonance frequency.

Solution:

$$f_R = \frac{1}{2\pi\sqrt{LC}}$$

$$f_R = \frac{1}{2 \times 3.14 \times \sqrt{0.0154 \times 0.0000161}}$$

$$f_R = \frac{1}{2 \times 3.14 \times \sqrt{.00000002479}}$$

FIGURE 24–1 Series resonant circuit.

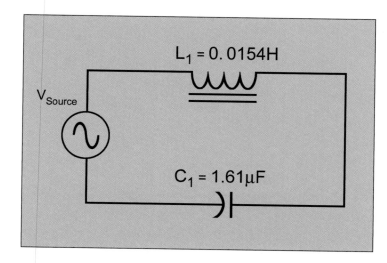

$L_1 = 0.0154H$

V_{Source}

$C_1 = 1.61\mu F$

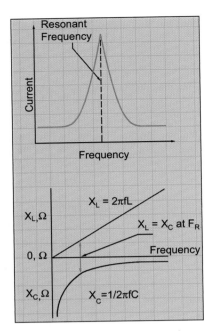

FIGURE 24–2 Current increase at resonant frequency.

$$f_R = \frac{1}{2 \times 3.14 \times 0.000157}$$

$$f_R = \frac{1}{.000989}$$

$$f_R = 1011.4 \text{ Hz}$$

In this example, the circuit will reach resonance at 1011.4 Hz. This is when both the inductor and the capacitor produce equal reactances. It is at this point that the two reactances are equal and 180° out of phase. Being equal and 180° out of phase with each other, they cancel each other out, and the total impedance is equal to zero.

At frequencies other than resonance, the current flow is limited by the net reactance, that is, the net effect of the inductive and capacitive reactances. If the net reactance is zero at resonance, then the impedance is zero at resonance. It follows, then, that at any frequency other than the resonant frequency (whether the frequency is above or below the resonant frequency), there would be some net reactance and some impedance. According to the formula, at a lower frequency, inductive reactance decreases, capacitive reactance increases, and total impedance increases. If the frequency were increased above the resonant frequency, inductive reactance would increase, capacitive reactance would decrease, and total impedance would increase.

RESONANCE CURRENT EFFECTS

The other issue is the effect of resonance on current. When the circuit reaches resonance, the current will suddenly increase because the only current-limiting factor will be the resistance of the wire in the coil. Figure 24–2 shows the effects of resonance on current in a series circuit.

INDUCTOR (Q) FACTOR

So far, to simplify the calculations, the discussion has assumed that an inductor has no resistance and that inductive reactance and capacitive reactance are the only current-limiting factors. Since inductors are coils of wire, the wire has some internal resistance. The amount of resistance compared to the inductive reactance determines the quality (Q) of the coil. Note that capacitors will also have some resistance; however, it is usually so small that it can be totally ignored.

24.2 Q Calculations: Method 1

The higher the ratio of inductive reactance to resistance, the higher the quality. An inductor constructed with a large wire will have a low wire resistance and a higher Q. Inductors with many turns and smaller wire will have a higher resistance and a lower Q. The formula for determining the Q of a coil is

$$Q = \frac{X_L}{R}$$

EXAMPLE 2

What is the Q of a circuit with a coil that has an inductance of 0.058 H at 400 Hz and a resistance of 15 ohms?

Solution:

$$X_L = 2\pi fL$$
$$X_L = 2 \times 3.14 \times 400 \times .058$$
$$X_L = 145.7$$

Then

$$Q = \frac{X_L}{R}$$

$$Q = \frac{145.7}{15} = 9.17$$

9.71

If the Q is greater than 10, the amount of resistance is considered negligible. Therefore, an inductor with Q of 10 or more is considered a pure inductor.

24.3 Q Calculations: Method 2

Since at resonance the current is greatest, the voltage measured across the coil can be considered the voltage output of the circuit at resonance. The voltage input is considered the voltage generated by the source. Thus, the formula is

$$Q = \frac{V_{out}}{V_{in}}$$

For example, if the source voltage in Figure 24–3 is 12 V and the voltage across L_1 is 96 V at resonance, the Q would be 96 V/12 V = 8. If you knew that Q was 8, then the voltage across the coil (L_1) could be calculated as the load voltage. This formula is expressed as

$$V_L = Q \times V_{source}$$

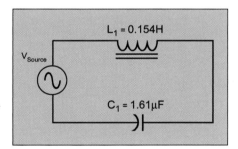

FIGURE 24–3   Series resonant circuit.

■ EXERCISES WITH SERIES RLC CIRCUITS

Use the circuit in Figure 24–4 as a reference for the following exercises. Solve for f_R X_L, X_C, Z, I_T, E_R, E_L, E_C, Q, cos θ, P_A, and P_T.

f_R:

$$f_R = \frac{1}{2\pi\sqrt{LC}}$$

$$f_R = \frac{1}{2 \times 3.14 \times \sqrt{(6 \times 10^{-5}) \times (90 \times 10^{-12})}}$$

$$f_R = \frac{1}{2 \times 3.14 \times \sqrt{5.4 \times 10^{-15}}}$$

FIGURE 24–4   Series RLC circuit.

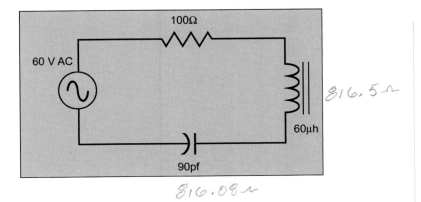

$$f_R = \frac{1}{4.61 \times 10^{-7}}$$

$$f_R = 2.166923 \text{ MHz}$$

X_L:

$$X_L = 2\pi fL$$

$$X_L = 2 \times 3.14 \times (2{,}166{,}923) \times (60 \times 10^{-6})$$

$$X_L = 816.5 \ \Omega$$

X_C:

$$X_C = \frac{1}{2\pi fC}$$

$$X_C = \frac{1}{2 \times 3.14 \times (2{,}166{,}923) \times (90 \times 10^{-12})}$$

$$X_C = 816.5 \ \Omega$$

Z:

$$Z = \sqrt{R^2 + (X_L - X_C)^2}$$

$$Z = \sqrt{100^2 + (0)^2}$$

$$Z = 100 \ \Omega$$

I_T:

$$I_T = \frac{E}{Z}$$

$$I_T = \frac{60}{100} = 0.6 \text{ A}$$

E_R:

$$E_R = I_R \times R$$

$$E_R = 0.6 \text{ A} \times 100 \ \Omega$$

$$E_R = 60 \text{ V}$$

E_L:

$$E_L = I_L \times X_L$$

$$E_L = .6 \times 816.5$$

$$E_L = 489.9 \text{ V}$$

E_C:

$$E_C = I_C \times X_C$$
$$E_C = .6 \times 816.5$$
$$E_C = 489.9 \text{ V}$$

Q:

$$Q = \frac{X_L}{R}$$
$$Q = \frac{816.5}{100}$$
$$Q = 8.165$$

$\cos \theta$:

$$\cos \theta = \frac{R}{Z}$$
$$\cos \theta = \frac{100}{100}$$
$$\cos \theta = 1, \theta = 0$$

P_A:

$$P_A = E_{app} \times I$$
$$P_A = 60 \times .6$$
$$P_A = 36 \text{ W}$$

P_T:

$$P_T = I^2 \times R$$
$$P_T = .6^2 \times 100$$
$$P_T = 36 \text{ W}$$
$$PF = \frac{P_T}{P_A}$$
$$PF = \frac{36}{36} = 1$$
$$PF = 100\%$$

24.4 Bandwidth

Bandwidth is a frequency measurement. At resonant frequency, the maximum current flows in a series RLC circuit. However, there are other frequencies that are close to this f_R. These frequencies cause much more current to flow in the circuit but less than the resonant frequency. Thus, any resonant frequency has an associated band or grouping of frequencies that cause resonant-like effects.

Look at Figure 24–5. The bandwidth, or spread, of these frequencies is determined by the Q of the resonant circuit and is set at the limit of the frequencies that cause a current that is at .707 of the maximum (resonant) current value. Bandwidth is the range of frequencies in

FIGURE 24–5   Determining Q.

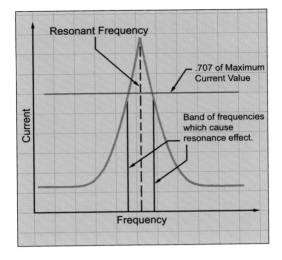

which the current remains at or above 70.7% of the current at resonance. The bandwidth is relative to the Q of the circuit and can be calculated using the following formula:

$$B = \frac{f_R}{Q}$$

Note that a lower Q will produce a wider bandwidth than a higher Q.

Using the results from the exercises and the circuit in Figure 25–4, calculate the f_R bandwidth:

$$B = 2.17 \text{ MHz}/8.15$$

$$B = 0.266 \text{ MHz}$$

■ VOLTAGE IN SERIES RLC CIRCUITS

Look at Figure 24–6. Remember that the V_L and V_C will be 180° out of phase with each other and that V_R will fall 90° between both. At resonance, the voltage across the inductor and capacitor will be greatest and 180° out of phase. If the output to the circuit were taken across either of these series components, voltage output would be multiplied by the Q factor of the circuit.

FIGURE 24–6   Voltage sine waves for RLC circuits.

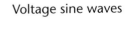

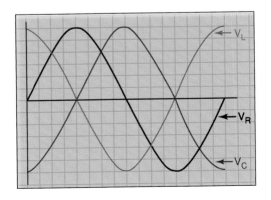

SUMMARY

For series RLC resonant circuits, inductance is measured in a unit called henrys (H) and symbolized by the letter "L." X_L symbolizes inductive reactance. All inductors contain some resistance.

The Q of an inductor is the ratio of inductive reactance to resistance. Capacitors have reactance symbolized by X_C. At resonance, X_C and X_L cancel each other out.

When a circuit reaches resonance, current reaches its maximum value. Voltage is maximum across each individual series inductor or capacitor and is 180° out of phase between the inductor and capacitor. The Q of the circuit can be calculated by dividing the voltage across the series capacitor or series inductor by the source voltage. Impedance is minimum at f_R and is equal only to the wire resistance of the inductor's coil.

Bandwidth is the range of frequencies in which the current remains at or above 70.0% of the current at resonance and is proportional to the Q of the circuit. The formula for bandwidth is $B = f_R/Q$.

REVIEW QUESTIONS

1. What is the relationship between capacitive reactance and inductive reactance at resonance?
2. What limits the current flow in a series RLC circuit at resonance?
3. Describe bandwidth in your own words.
4. How does the Q of the circuit affect bandwidth?
5. How is the term *bandwidth* as used in this chapter related to the term *bandwidth* that you often hear used with respect to the Internet?
6. What happens to the following quantities in a series RLC circuit at resonance?

   a. Impedance
   b. Current flow
   c. Inductive reactance
   d. Capacitive reactance
   e. Circuit voltage across
      i. The inductor
      ii. The capacitor
      iii. The whole circuit

PRACTICE PROBLEMS

All the practice problems refer to the following figure:

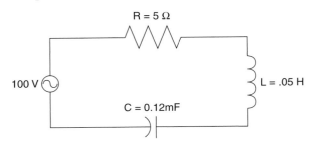

R = 5 Ω

100 V

L = .05 H

C = 0.12mF

1. Find the resonant frequency.
2. Find I_T at f_R.
3. What is the Q of the circuit?
4. What is the bandwidth of this circuit?
5. What are the frequencies in the bandwidth?

chapter 25

Parallel RLC Circuits and Resonance

■ OUTLINE

■ OVERVIEW

Resonance occurs at a frequency when X_L is equal to X_C. Remember that resonance is defined as the frequency at which the inductive reactance (X_L) is equal to the capacitive reactance (X_C). This chapter provides an excellent opportunity to review this concept. The formula for resonance is

$$2\pi f L = \frac{1}{2\pi f C}$$

$$f^2 = \frac{1}{4\pi^2 LC}$$

$$f_R = \frac{1}{2\pi\sqrt{LC}}$$

Remember that voltage is the same across every leg, or branch, of a parallel circuit and is used as the reference. This chapter is organized in much the same way as chapter 24 and provides a substantial amount of practice in the analysis of parallel resonant circuits.

■ OBJECTIVES

After completing this chapter, you should be able to:

1. Describe the behavior of a parallel resonant circuit.
2. Calculate for unknown parameters in a parallel resonant circuit.

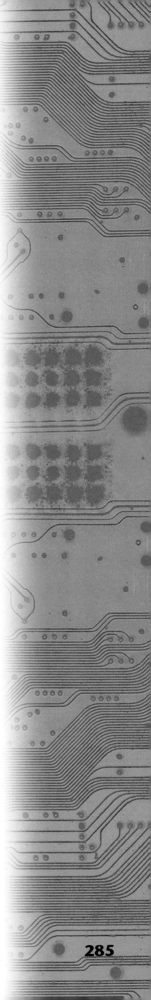

FIGURE 25–1   LC tank circuit.

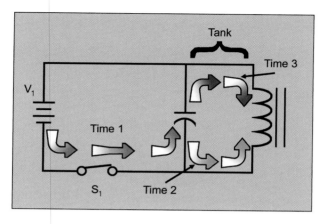

■ PARALLEL CIRCUIT RESONANCE

25.1 Basic Parallel LC Operation

In a parallel circuit, inductive current and capacitive current subtract from each other because they are 180° out of phase with each other. This means that there will be a minimum of line current at the point of resonance.

Look at Figure 25–1. At time 1, the switch (S_1) is closed, and the voltage from V_1 charges the capacitor. At time 2, the capacitor is fully charged, and the switch is opened. Now the only discharge path for the capacitor is through the inductor. As the capacitor discharges through the inductor, the inductor's magnetic field expands, storing the charge. Once the inductor is charged (time 3), it will discharge back into the capacitor, starting the cycle all over again. In theory, the cycle will continue forever without any resistance in the tank circuit.

25.2 Ideal Versus Real

As shown in Figure 25–1, with pure inductance and capacitance, there is no true power consumed. For this reason, the oscillations at resonance will go on forever at the voltage level or amplitude applied by the battery power supply (see Figure 25–2).

In reality, a parallel resonant circuit has resistance. Resistance will consume power. This power consumption is the I^2R loss in the wire. The effect of the loss appears progressively as each swing gets progressively smaller and stops after a few cycles. This is represented by the sine wave in Figure 25–3.

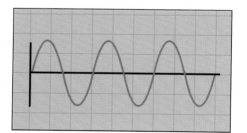

FIGURE 25–2   Ideal circuit oscillation continues forever.

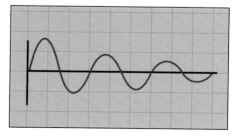

FIGURE 25–3   Real circuit oscillations (damped).

■ RESISTANCE IN PARALLEL LC CIRCUITS

In "real" parallel circuits, the resistance of the circuit is equal to the resistance of the wire used in the winding of the inductor. The resistance is in series with the inductive reactance. The resistance in the capacitor is very small and can usually be ignored for these calculations.

Because the resistance is in series with the inductor when $X_L = X_C$, the current flow through the inductive leg is going to be slightly less than the current in the capacitive leg. Therefore, the two currents do

FIGURE 25–4 Parallel RLC resonant circuit.

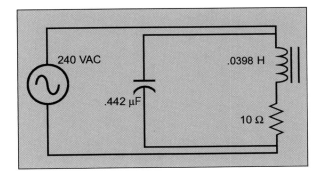

not exactly cancel each other out. In the real parallel circuit, this means that the capacitive current is greater than the inductive current and that the line current will lead the applied voltage. For the reactive currents to actually cancel each other out and the circuit to act purely resistive, a frequency slightly lower than the defined frequency is needed. At this point, $\theta = 0°$.

25.3 Practical Circuit Example

In Figure 25–4, the inductor has an inductance of .0398 H, and the wire resistance is 10 Ω. The capacitor has a capacitance of .442 μF. This circuit will reach resonance at 1,200 Hz. At that time, the inductor and the capacitor will exhibit a reactance of 300 Ω each.

The calculations for the values in a parallel resonant circuit are a bit more complex than calculations for series resonant circuits. Theoretically, when a parallel circuit reaches resonance, the total circuit current should reach zero, and the total circuit impedance should become infinite because the capacitive and inductive currents cancel each other out.

25.4 Circuit Calculations

In the circuit shown in Figure 25–4, solve for the following: f_R, X_L, X_C, I_C, Z_{LR} (impedance of the inductor and the resistance of the coil windings), I_{LR}, tan θ, I_{circ} (current inside the capacitor-inductor parallel connections), Q, I_{line}, (total circuit current), bandwidth (BW), f_1 (minimum frequency related to bandwidth), f_2 (maximum frequency related to bandwidth), and Z.

f_R:

$$f_R = \frac{1}{2\pi\sqrt{LC}}$$

$$f_R = \frac{1}{6.28\sqrt{(3.98 \times 10^{-2}) \times (4.42 \times 10^{-7})}}$$

$$f_R = 1{,}201 \text{ Hz}$$

X_L:

$$X_L = 2\pi fL$$
$$X_L = 6.28 \times 1{,}201 \times .0398$$
$$X_L = 300 \ \Omega$$

X_C:

$$X_C = X_L = 300 \ \Omega$$

I_C:

$$I_C = \frac{E}{X_C}$$
$$I_C = \frac{240}{300}$$
$$I_C = .8 \ A$$

Z_{LR}:

$$Z_{LR} = \sqrt{X_L^2 + R^2}$$
$$Z_{LR} = \sqrt{300^2 + 10^2}$$
$$Z_{LR} = 300.17 \ \Omega$$

I_{LR}:

$$I_{LR} = \frac{E_A}{Z_{LR}}$$
$$I_{LR} = \frac{240}{300.17}$$
$$I_{LR} = 0.7995 \ A \approx 0.8 \ A$$

$\tan \theta$:

$$\tan \theta = \frac{X_L}{R}$$
$$\tan \theta = \frac{300}{10}$$
$$\tan \theta = 30$$
$$\theta = 88°$$

I_{circ}:

$$I_{circ} = I_C = .8 \ A \ \text{(remember that the circuit is at resonance)}$$

Q:

$$Q = \frac{X_L}{R} \text{ or } Q = \frac{I_{circ}}{I_{line}}$$
$$Q = \frac{300}{10}$$
$$Q = 30$$

I_{line}:

$$I_{line} = \frac{I_{circ}}{Q}$$
$$I_{line} = \frac{.8 \ A}{30}$$
$$I_{line} = .0267 \ A$$

or

$$I_{line} = \sqrt{I_{LH}^2 + (I_C - I_{LV})^2}$$

where I_{LH} = horizontal component of the I_{LR} current:

$$I_{LH} = \cos 88° \times .8$$
$$I_{LH} = .027920$$

and I_{LV} = vertical component of the I_{LR} current:

$$I_{LV} = \sin 88° \times .8$$
$$I_{LV} = .7995$$
$$I_{line} = \sqrt{I_{LH}^2 + (I_C - I_{LV})^2}$$
$$I_{line} = \sqrt{(.0279)^2 + (.8 - .7995)^2}$$
$$I_{line} = \sqrt{.00078 + .000001}$$
$$I_{line} = .0279 \text{ A}$$

Notice that the values are close but not quite the same (.0001). The current calculations are more accurate than the Q value calculations for determining I_{line}. But both will give good values for circuit analysis.

BW:

$$BW = \frac{f_R}{Q}$$
$$BW = \frac{1,201}{30}$$
$$BW = 40$$

f_1:

$$f_1 = f_R - \frac{BW}{2}$$
$$f_1 = 1,201 - \frac{40}{2}$$
$$f_1 = 1,201 - 20$$
$$f_1 = 1,181$$

f_2:

$$f_2 = f_R + \frac{BW}{2}$$
$$f_2 = 1,201 + \frac{40}{2}$$
$$f_2 = 1,201 + 20$$
$$f_2 = 1,221$$

Z:

$$Z = \frac{E_A}{I_{line}}$$
$$Z = \frac{240}{.0279}$$
$$Z = 8,602 \ \Omega$$

Table 25.1

L, in μH	C, in pF	Freq, in kHz
239	424	500
239	212	707
239	106	1,000
239	53	1,410
239	26.5	2,000

■ PARALLEL RESONANT CIRCUIT TUNING

When tuning a parallel resonant circuit, the operator usually tunes to the point of the least amount of line current. The parallel resonant circuit has its maximum impedance value when $X_L = X_C$. Tuning below resonance results in X_L decreasing toward zero reactance when the frequency is 0 Hz. This means that an undefined current will flow through the inductive leg at this point. Increasing the frequency toward an undefined value results in X_C approaching zero reactance and an undefined current flow in the capacitive leg. Table 25–1 shows the relationship between frequency and the LC components of a circuit. In the table, the LC circuit is tuned by varying the capacitor.

Notice the relationship between the reduction in pF and the increase in frequency. For example, the pF was reduced by ¼, and the frequency doubled.

■ SUMMARY

The RLC circuit is often called a tank circuit. When the parallel circuit reaches resonance, the current drops, and the total impedance increases. When the circuit is at the point of resonance, the total current is determined by the amount of pure resistance in the circuit. Total circuit current and total impedance in a tank circuit are proportional to the Q of the circuit.

■ REVIEW QUESTIONS

1. What is the relationship between capacitive reactance and inductive reactance at resonance?
2. What limits the current flow in a parallel RLC circuit at resonance?
3. Describe bandwidth in your own words.
4. How does the Q of the circuit affect bandwidth?
5. What happens to the following quantities in a parallel RLC circuit at resonance?
   a. Impedance
   b. Current flow
   c. Inductive reactance
   d. Capacitive reactance
   e. Circuit voltage across
      i. The inductor
      ii. The capacitor
      iii. The whole circuit

PRACTICE PROBLEMS

All the practice problems refer to the following figure:

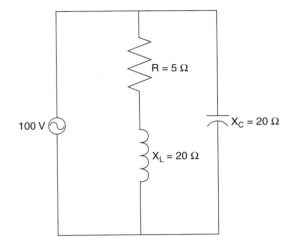

1. Find I_T.
2. Does the current lead or lag the applied voltage?
3. Find the circulating current.
4. Find the Q of the circuit.
5. Prove your calculation of Q using $Q = \dfrac{I_{tank}}{I_{line}}$

PART

6

ADDITIONAL AC TOPICS

chapter 26

Use of Filters to Control AC Signals

■ OUTLINE

■ OVERVIEW

In previous chapters, you learned about resonance, inductive reactance, and capacitive reactance. These characteristics are used to design useful circuits that will allow current or voltage at certain frequencies to pass to the load, while at other frequencies current and voltage are reduced at the load. These special circuits are called *filters* or *wave traps*.

It is a filter that distinguishes one radio or television station from another. It is also the filter that prevents one radio station from interfering with another. It is not unusual to have frequencies with considerable power generated on an AC power line because of operating various types of loads. These frequencies can cause interference with radios, televisions, computers, or other electronic devices operating on the same power line. The most common types of interference are static or a low hum. Power line filters can be installed to prevent these types of interference.

In this chapter, you will learn about the various types of filters and how they work. This information is used in the design of a wide variety of filters ranging from radio and television circuitry to harmonic filters used in a broad range of power system applications.

■ OBJECTIVES

After completing this chapter, you should be able to:

1. Explain the theory of operation of common type filters.
2. Draw the pass bands for various filters.
3. Design very simple filter circuits.

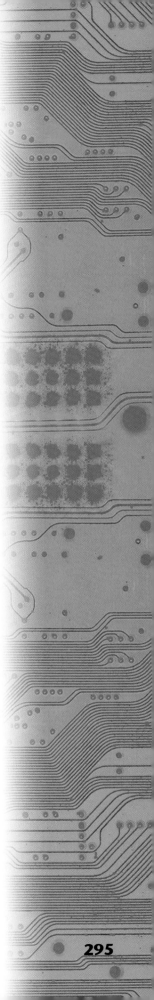

■ FILTER OPERATION AND CLASSIFICATION

26.1 Filter Operation

Figure 26–1 shows the basic concept of filtering. Note that with different frequency components coming into the filter, only one frequency component is allowed to leave. The low-pass filter blocks the high frequency (100 kHz) and passes the low frequency (1 kHz). The high-pass filter works in an opposite manner.

26.2 Filter Classification

There are many types of electrical filters used for different applications:

- Direct current combined with alternating current
- Transformer coupling
- Capacitive coupling
- Bypass capacitors
- Filter circuits
- Low-pass filters
- High-pass filters
- Resonant filters
- Interference filters
- Band-pass
- Band-reject

This chapter discusses the four most common filters:

- Low-pass
- High-pass
- Band-pass
- Band-reject

FIGURE 26–1 Low- and high-pass filter operation.

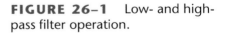

Band-pass and band-reject filters are special combinations of low- and high-pass filters.

■ LOW-PASS FILTERS

26.3 Low-Pass Filter Operation

A simple low-pass filter uses an inductor in series with a load resistor. A low-pass filter is shown in Figure 26–2. At low frequencies, the inductive reactance will be low, and most of the source voltage will be dropped across the load (R_L). Remember that $X_L = 2\pi fL$.

As the frequency increases, X_L increases. The result of the increase in X_L is that more voltage is dropped across the inductor and less voltage is dropped across the load. When the X_L value (measured in ohms) is 10 times greater than the load resistance (also measured in ohms), there is virtually no signal across the load.

The opposite is also true. If the value of X_L is 10 times less than the value of the resistance, most of the signal will reach the load. Typically, low-pass filters are designed to deliver most of the signal to the load at lower frequencies.

The half-power point is defined as the point at which the load resistance is dissipating half the maximum value. It is easy to assume that this will be true when X_L equals the value of the resistance, and in fact this is the case.

To prove it, start by considering the fact that the maximum power will be dissipated by the resistor when the voltage on the resistor is equal to the supply voltage. This is true because at this point the voltage is maximum, and therefore the power dissipation is maximum. This will occur at the maximum voltage drop across the load resistor, which is represented as E_{sup}. Now the formula for the maximum power is given by

$$P_{max} = \frac{E_{sup}^2}{R_L}$$

FIGURE 26–2   Low-pass filter.

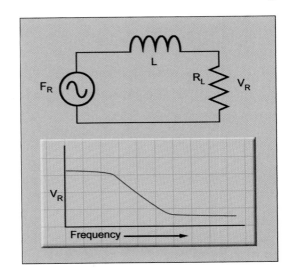

Solving for E_{sup} gives

$$E_{sup}^2 = P_{max} \times R_L$$

and

$$E_{sup} = \sqrt{P_{max} \times R_L}$$

The voltage for one-half power is given by

$$E_{\frac{1}{2}} = \sqrt{\frac{P_{max}}{2} \times R_L}$$

Substituting for P_{max} yields

$$E_{\frac{1}{2}} = \sqrt{\frac{\left(\dfrac{E_{sup}^2}{R_L}\right)}{2} \times R_L} = \sqrt{\frac{E_{sup}^2}{2}}$$

The voltage drop across the resistor can also be calculated using the voltage divider formula:

$$E_{\frac{1}{2}} = \frac{R_L}{\sqrt{X_L^2 + R_L^2}} \times E_{sup}$$

This means that

$$\sqrt{\frac{E_{sup}^2}{2}} = \frac{R_L}{\sqrt{X_L^2 + R_L^2}} \times E_{sup}$$

Squaring both sides yields

$$\frac{E_{sup}^2}{2} = \frac{R_L^2}{X_L^2 + R_L^2} \times E_{sup}^2$$

Dividing both sides by E_{sup}^2 gives

$$\frac{1}{2} = \frac{R_L^2}{X_L^2 + R_L^2}$$

Cross-multiplying gives;

$$X_L^2 + R_L^2 = 2 \times R_L^2$$

And subtracting R_L^2 from both sides yields

$$X_L^2 = R_L^2$$

Taking the square root of both sides yields

$$X_L = R_L$$

Notice that this will be true even if X_L is negative. Of course, inductive reactance cannot be negative; however, capacitive reactance can be, so the same formula holds for a series RC circuit.

At half power, the voltage across the load and the current flowing through the load is .707 times the maximum possible value. All frequencies less than one-half maximum power are considered to pass to the load. All frequencies greater than one-half power are rejected and never reach the load.

Note that the signal at the load resistor first starts to degrade when X_L is ⅒ the value of the resistance.

26.4 Inductive Low-Pass Filter

As an example, design a low-pass filter that will pass all frequencies up to 400 Hz to a load of 600 Ω. The first step is to determine the value of the inductor. Since the half-power point occurs when $X_L = R_L$, the inductor will equal

$$L = \frac{X_L}{2\pi f}$$

$$L = \frac{600\ \Omega}{6.28 \times 400\ \text{Hz}}$$

$$L = .239\ \text{H}$$

26.5 Capacitive Low-Pass Filter

Another low-pass filter is shown in Figure 26–3. This filter has a capacitor in parallel with the load resistor. At low frequencies, the capacitor will exhibit a high reactance, and most of the current will flow through the load resistor.

As the frequency is increased, the capacitive branch will draw greater current because of the decreasing value of X_C. As a result, greater current will flow through the series resistor, causing the voltage across it to increase, with the voltage across the filter and the load decreasing. This type of filtering is often used across the power line to reduce high-frequency interference.

FIGURE 26–3   Capacitive low-pass filter.

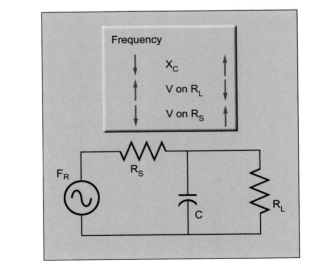

FIGURE 26–4   Combined LC
low-pass filter.

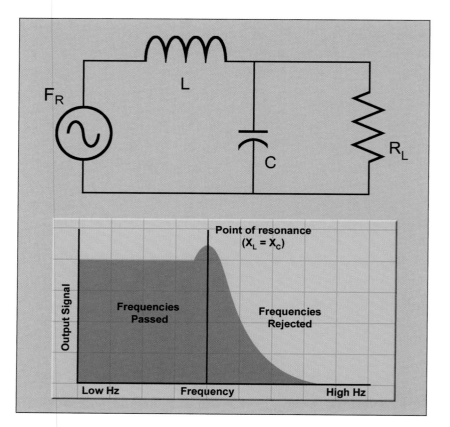

26.6 Combined LC Low-Pass Filter

Another example of a low-pass filter is shown in Figure 26–4. This filter consists of an inductor in series with a capacitor in parallel with the load. At the lower frequencies, very little voltage will be dropped across the inductor. Most of the source voltage will be across the very high reactance of the capacitor and the load. As the frequency is increased, more of the source voltage will be across the inductor, and the voltage across the capacitor and the load will decrease. This filter has a sharper slope and is more efficient than the filter in Figure 26–3 because both the inductor and the capacitor serve to reduce the voltage across the load as the frequency increases.

When the circuit reaches the resonant frequency ($X_L = X_C$), the output signal will peak. This will be followed by a sharp drop in the output signal as the frequency continues to rise.

■ HIGH-PASS FILTERS

26.7 High-Pass Filter Operation

Figure 26–5 shows a simple high-pass filter that consists of a capacitor in series with the load. This configuration is easily recognized as a simple series RC circuit.

At low frequencies, the X_C is very high, and the capacitor will drop the most voltage. As the frequency is increased, less voltage will be

FIGURE 26–5 High-pass filter.

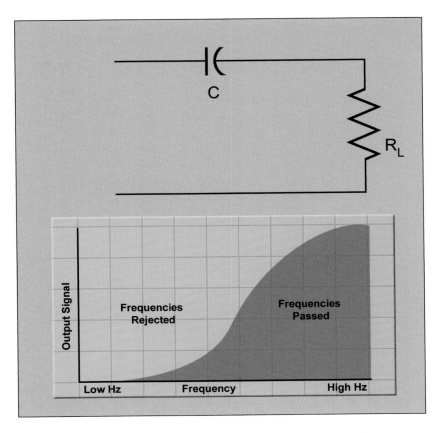

dropped across the capacitor, and more voltage will be dropped across the load. This type of filter can be used to eliminate interference (at power line frequencies) from entering the load. Some intercom systems use the power line to carry the signal, and this filter would block these low frequencies.

As you saw previously, the half-power point for this filter will occur when the capacitance reactance (X_C) equals the load resistance:

$$X_C = R_L$$

26.8 Capacitive High-Pass Filter

When the X_C value is ⅒ as great as the load resistance, there is virtually no voltage signal across the capacitor. As an example, consider a filter that will pass all frequencies above 400 Hz to a 600-Ω load. The first step is to calculate the value of C.

The capacitance value equals

$$C = \frac{1}{2\pi f X_C}$$

$$C = \frac{1}{6.28 \times 400 \times 600}$$

$$C = \frac{1}{1,507,200}$$

$$C = 0.66 \ \mu F$$

FIGURE 26–6   Inductive high-pass filter.

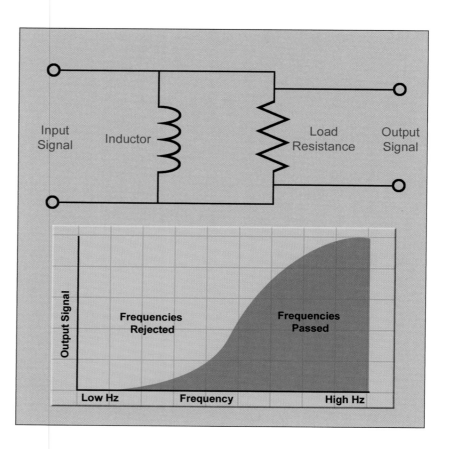

26.9 Inductive High-Pass Filter

In Figure 26–6, an inductor is in parallel with the load resistor. At low frequencies, the inductive reactance (X_L) is low, allowing most of the circuit current to flow through the inductor. As the frequency increases and the inductive reactance (X_L) is also increasing, more of the current flows through the load. When the value of the inductive reactance (X_L) becomes 10 times the value of the load resistor, all the signal flow is through the load resistor, and the inductor no longer has any effect on the circuit.

26.10 Combined LC High-Pass Filter

In Figure 26–7, the series resistor is replaced with a capacitor. Again, at the low frequencies, more of the source voltage will be across the capacitor (because of its high reactance), and less voltage will be across the inductor and the load (because of the low reactance of the inductor). As frequency is increased, the capacitive reactance decreases, and the inductive reactance increases. The voltage across the load increases with the increasing inductive reactance.

■ BAND-PASS FILTERS

26.11 Band-Pass Filter Operation

As learned in previous chapters, a tank circuit exists when a circuit contains an inductor and capacitor in parallel. In the tank circuit, the inductor and capacitor currents are 180° out of phase when these two

FIGURE 26-7 Combined LC high-pass filter.

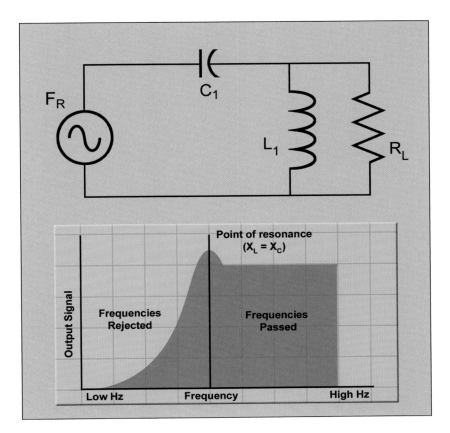

components are connected in parallel and the voltages are in phase. When the parallel LC circuit is at resonance, the capacitive current and the inductive current cancel each other out because of this phase relationship. Since the inductive current and the capacitive current cancel each other out, the total circuit current is at a minimum value, and the impedance is at the maximum value. This cancellation takes place at the resonant frequency f_R and can be controlled by carefully selecting the appropriate capacitor and inductor for the tank circuit.

One of the primary uses for this type of tank circuit is to pass a specific frequency based on the resonant frequency of the newly designed circuit. At frequencies other than the resonant frequency, a current will flow through the tank circuit because the complete cancellation of the current flows from the inductor and capacitor will not take place. Remember that other than at resonance, the X_L and the X_C values will always be unequal; therefore, the current flows from each will also be unequal. At frequencies other than the resonant frequency, the signal is shunted away from the load and through the tank circuit.

The concept of resonance is used to pass a selected band of frequencies and to reject all others. The width of the band is determined by the value of the inductor and capacitor. Remember also that the bandwidth is defined as those frequencies at or above the half-power level (.707 V). Figure 26–8 shows a band-pass filter that consists of a series LCR circuit with high-Q components. The resistance in this circuit is considered to be only the load.

FIGURE 26–8   Band-pass filter with high-Q components.

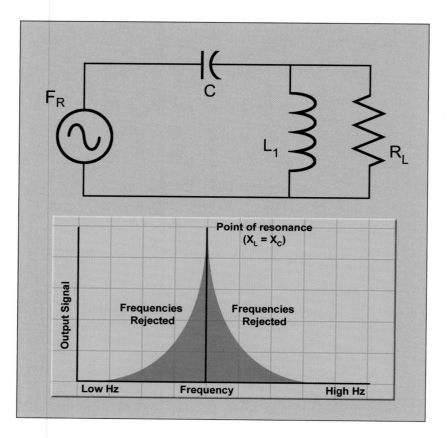

The resonant frequency (F_R or F_0) for this circuit is determined by

$$f_R = \frac{1}{2\pi\sqrt{LC}}$$

To determine the bandwidth, the Q of the circuit must first be determined:

$$Q = \frac{X_L}{R}$$

$$\text{BW} = \frac{f_R}{Q}$$

$$f_1 = f_R - \frac{\text{BW}}{2}$$

$$f_2 = f_R + \frac{\text{BW}}{2}$$

Note that f_1 and f_2 are the minimum and maximum frequencies for the bandwidth, respectively.

A general rule of thumb relative to the Q value of the components is that a circuit with high-Q components equals a small, or narrow, bandwidth. A circuit with low-Q components equals a wide bandwidth.

26.12 Band-Pass Filter with Parallel Load

Figure 26–9 is a configuration for the band-pass filter with low-Q components. In this example, the load (R_L) is across the resonant circuit. This parallel combination is in series with a resistor connected to the

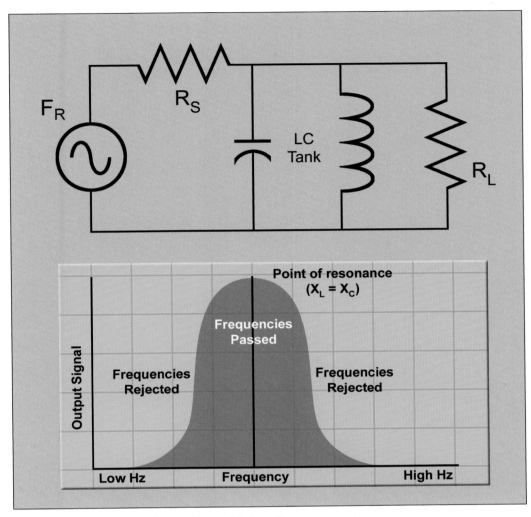

FIGURE 26-9   Band-pass filter with low-Q components.

source. At resonance, minimum line current occurs because of the cancellation of the inductive and capacitive currents. At this point, the series resistor will drop a minimum voltage, and the maximum voltage will be dropped across the load resistor. This voltage drop is due to the LC tank having a very high impedance at resonance, so the parallel R_L will see a high voltage across the inductor.

■ BAND-REJECT FILTERS

26.13 Band-Reject Filter Operation

Figure 26–10 is an example of a band-reject filter with high-Q components. This circuit consists of an inductor and capacitor connected in parallel with each other and in series with the load resistor. (Notice that to go from band-pass to band-reject filtering, the load was moved from the parallel to the series position with the LC tank portion of the circuit.) At resonance, the impedance of the resonant circuit will be at maximum, and a minimum current will flow to the load. This is due to the inductive and capacitive currents canceling

FIGURE 26–10   Band-pass filter with high-Q components.

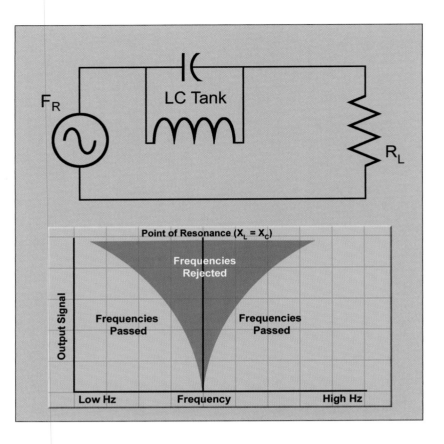

each other. Off resonance, a greater current flow will result, and the voltage across the load increases.

26.14 Band-Reject Filter with Parallel Load

Figure 26–11 is an example of a band-reject filter circuit with low-Q components. In this case, the load is connected in parallel to the resonant series LC circuit. A resistor (R_S) is connected to this configuration in series with the source. At resonance, the inductive reactance and the capacitive reactance cancel each other because their currents are 180° out of phase. This places the load resistor (R_L) in parallel with a very low impedance. The current flow through the series resistor (R_S) will be maximum at this point, and this will produce a minimum voltage across the load resistor.

■ COMPLEX FILTERS

26.15 T-Type Filter

More often than not, filters are more complex than the ones in the previous examples. Figures 26–12 and 26–13 show two T-type filters. The filter in Figure 26–12 is a high-pass filter, and the one in Figure 26–13 is a low-pass filter. Several sections of T-type filters may be joined in some applications. These types of filters improve filtering by increasing sharpness of the cutoff frequencies.

FIGURE 26–11 Band-reject filter with low-Q components.

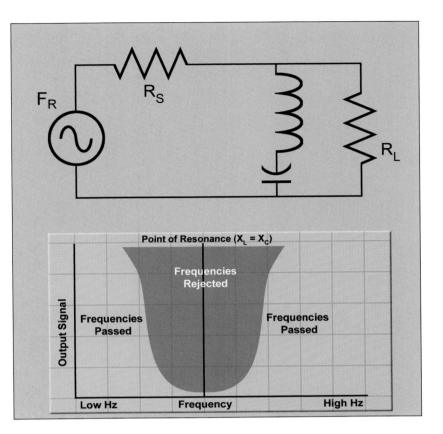

FIGURE 26–12 T-type high-pass filter.

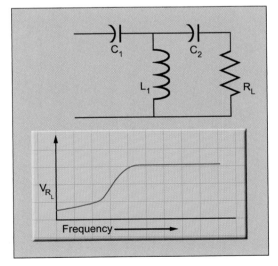

In Figure 26–12, the high frequencies are allowed to develop across the load by the inductor (L_1). At low frequencies, the inductor essentially acts as a shunt across the load.

In Figure 26–13, the low frequencies are allowed to pass by the inductors (L_1 and L_2), and the capacitor acts as a high impedance for the load.

FIGURE 26–13 T-type low-pass filter.

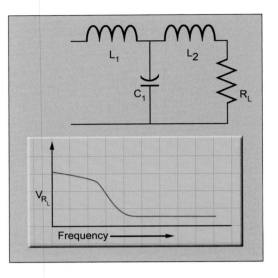

26.16 π-Type Filter

Figure 26–14 is an illustration of a π-type filter. This filter is commonly found in power supplies to smooth the rectified wave to pure DC. The input of the filter shown is the 120-Hz pulses of a full-wave rectifier supplied by a 60-Hz AC source. The capacitors usually have very large capacitance (10 mF to 1,000 mF). The inductance, at the same time, has a value in the range of 10 to 100 H. This combination presents a high opposition to the changing voltage and current of the rectified input. The capacitors charge to the peak value of the rectified voltage so that if the input is 100 V_{RMS}, the output will be 141.4 VDC. This filter would be classified as a low-pass filter.

FIGURE 26–14 π-type low-pass filter.

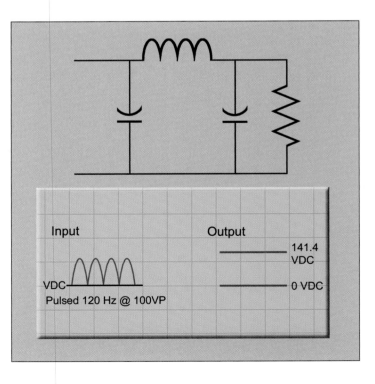

FIGURE 26–15 π-type high-pass filter.

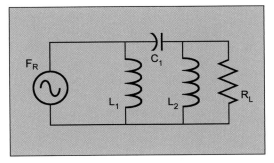

A high-pass π filter is shown in Figure 26–15. In this circuit, the inductors now form the legs of the pie, with the capacitor between them. Low frequencies will be blocked by the capacitor and shunted through the inductors around the load. The high-frequency components of the input will develop very little voltage across the series capacitor (C_1), allowing most of the voltage to be produced across the load (R_L).

26.17 Other Complex Filters

Even more complex filters are shown in Figures 26–16 and 26–17. Figure 26–16 is a complex band-pass filter, and Figure 26–17 is a complex band-reject filter.

FIGURE 26–16 Complex-type band-pass filter.

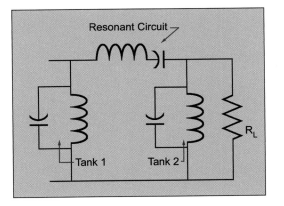

FIGURE 26–17 Complex-type band-reject filter.

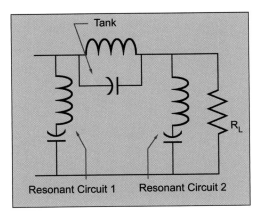

Notice that the band-pass filter has the resonant circuit components in series with the load (R_L) and that the band-reject filter has the tank circuit in series with the load (R_L). These filters can be tuned to pass a given band of frequencies and reject interfering frequencies operating very close to the desired band. These combinations may also be combined to form even more complex filters. Engineers specializing in this field are usually the designers of these types of filters.

■ SUMMARY

In general, characteristics for high-pass filters are the following:

- Coupling capacitance is in series with the load. The X_C can be low for high frequencies to be passed to the load, while low frequencies are blocked.
- Inductors are in parallel across the load. The shunt X_L can be high for high frequencies to prevent a short circuit across R_L, while low frequencies are bypassed.

Characteristics for low-pass filters are the following:

- Inductors are in series with the load. The high X_L for high frequencies can serve as a choke or block, while low frequencies can be passed to the load.

- A bypass capacitor is in parallel across the load. The high frequencies are bypassed by a small X_C, while low frequencies are not affected by the shunt path.

Filters are becoming increasingly important in electrical work. The security of companies (as well as government secrets) is in jeopardy when computers are radiating signals or information that escape through the power source. This information could be collected by anyone having access to a receptacle or breaker box. The information could then be used to compromise company and government secrets.

The needs of reducing interference between equipment and the maintenance of secure systems can be a source of increased income for the electrical industry.

■ REVIEW QUESTIONS

1. Discuss the behavior of low-pass, high-pass, band-pass, and band-reject filters.
2. What is the underlying concept that makes it possible to create filters using R, L, and C components?

3. How is the half-power point related to the voltage drop across the load resistor in a high-pass or a band-pass filter?
4. Discuss the applications for filters using low-Q components. High-Q components.

■ PRACTICE PROBLEMS

1. Name six types of filters.
2. For a low-pass RL filter:
   a. The load resistor will have a voltage across it when the frequency is _____ the half-power point.
   b. The half-power point is defined as the point where X_L is equal to _____.

   c. What is the cutoff frequency for a low-pass RL filter having a 3.3-k Ω resistor and a 1-mH coil?
3. For a low-pass RC filter:
   a. Low-pass RC filters have the capacitor connected in _____ with the load resistor.
   b. Design a 2-kHZ low-pass RC filter with a 1-k Ω series resistor; that is, find C.

4. For a low-pass LC filter:

    a. In a low-pass LC filter, the output signal will peak when $X_L =$ ____. (resonant frequency)

    b. What is the peak frequency for the speaker in the following figure?

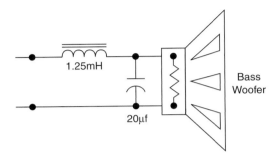

1.25mH

20µf

Bass Woofer

    c. Find the actual cutoff frequency (that frequency where the signal is at the upper half-power point).

5. For a high-pass RC filter:

    a. Using a .1-µf capacitor, find the resistance needed to make a high-pass 300-Hz filter.

6. For a high-pass RL filter:

    a. Find the cutoff frequency for a high-pass RL filter with a 22-K Ω resistor and a 500-µH coil.

7. For high-pass LC filter:

    a. The output of a high-pass LC filter increases as the X_L ____.

8. For a band-pass filter:

    a. What is the center frequency for a series band-pass filter with a 100-µH coil and an 800-pF capacitor?

    b. Find the center frequency for a parallel band-pass filter with a 100-µH coil and a 100-pF capacitor.

9. For a band-reject filter:

    a. Find the center frequency for a series band-reject filter having a 42.9-mH coil and a .0041-µf capacitance.

    b. If the Q of the circuit in the previous question is 3, what is the bandwidth?

    c. Find f_1 and f_2 for the previous circuit.

chapter 27

Power Factors in AC Circuits

OVERVIEW

Previous chapters have covered power factor when discussing RLC circuits. In this chapter, the subject is discussed in greater detail. Using characteristics that have already been associated with combinations of resistance, capacitance, and inductance, you will be able to determine the power factor of various circuits.

Previously, power was calculated by multiplying the voltage times current. Such an approach is sufficient for a beginning understanding of power, but in actual circuits it is an oversimplification. When current and voltage are in phase, the circuit true power is found by multiplying $E \times I$. In reality, all circuits have some amount of either inductance or capacitance. Since these circuits contain inductance or capacitance, their effect on the circuit—their X_L or X_C—changes the phase relationship.

In residential electrical installations, the reactance load is low, but in industrial applications, it can be significant. In circuits with both resistance and reactance, there are two types of power at the load: true power and reactive power. True power is measured in watts and reactive power in VARs (volt-amperes-reactive).

The apparent power of the circuit is measured in volt-amperes and is the vector sum of the true power and reactive power. Although reactive power does no work, it costs money. Reactive loads draw current, which means that the electric company must provide greater generator capacity.

OBJECTIVES

After completing this chapter, you should be able to:

1. Define the term *power factor* (PF) and become familiar with formulas for calculating PF.
2. Define and distinguish between the terms *watts*, *VARs*, and *volt-amperes*.

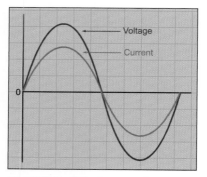

FIGURE 27–1 In-phase voltage and current.

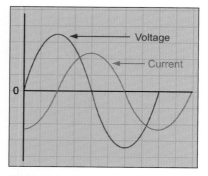

FIGURE 27–2 Current lagging voltage.

◼ TYPES OF POWER

27.1 True Power

When an AC voltage is applied to a resistor, the current wave shape will be a copy of the voltage; that is, it will rise and fall at the same rate and in the same direction as the voltage and will reverse in polarity at the same time that the voltage reverses polarity. In this condition, the current is said to be in phase with the voltage (see Figure 27–1).

This in-phase power component can be calculated by the E × I formula and is known as true power. It is also resistive power: power dissipated as heat.

27.2 Reactive Power

In an inductive circuit, the current changes at an exponential rate when the source voltage changes. The current lags the voltage across the inductor by 90° (see Figure 27–2) because the induced voltage across the inductor is 180° out of phase with the changes in source voltage.

As indicated, voltage and current are not in phase with each other. Because of circuit inductance, they will always be slightly out of phase. As you can see, the reactance does not dissipate power but stores energy in the electrostatic (capacitive) or magnetic (inductive) field. This stored power is returned to the circuit whenever E and *I* reach opposite polarity.

27.3 Apparent Power

Apparent power can be computed by multiplying the circuit voltage (V_{applied}) times the total current flow (I_{T}) with no regard for phase angle. Remember that total circuit current is determined by combining the reactive and the resistive components. The total current would be the measured current, as it is the amount of current that is apparently being used by the circuit and therefore is a measurable quantity. For instance, in a circuit with 220 V and total current of 14 A, the volt-amperes (VA) are

$$VA = V_{\text{applied}} \times I_{\text{T}}$$
$$VA = 220 \times 14$$
$$VA = 3,080$$

27.4 Power Factor

The power factor is not an angular measure but a numerical ratio. This is the ratio of true power to apparent power. If the ratio is 1, or unity, it means that the circuit is a resistive circuit. The opposite is also true: if the circuit has no resistive component and is entirely reactive, the power factor is 0. The power factor can be calculated by dividing the circuit resistive value by its similar total value. For example, power factor can be calculated by dividing the voltage drop across the resistor by the total circuit voltage, resistance by impedance, or watts by volt-amperes.

For example, given a true power of 1,936 W with a VA of 3,080, find the power factor (PF). To turn the decimal into a percentage, multiply the answer by 100:

$$PF = \frac{true\ power}{apparent\ power}$$

$$PF = \frac{W}{VA} \times 100$$

$$PF = \frac{1,936}{3,080} \times 100$$

$$PF = 62.9\%$$

or

$PF = V_R$ = voltage drop across the resistor, and V_T = the total voltage applied to the circuit

or

$$Series\ circuit: PF = \frac{Z}{R}$$

$$Parallel\ circuit: PF = \frac{R}{Z}$$

27.5 Angle Theta

The angular displacement of the voltage and current that are out of phase with each other is called *angle theta*. Since the power factor is the ratio of true power to apparent power, the phase angle of voltage and current is formed between the resistive leg and the hypotenuse. The cosine of theta is equal to the watts divided by the volt-amps. In other words, the cosine of theta is equal to the power factor.

Because it indicates the resistive component, cos θ is the power factor of the circuit. Figure 27–3 shows the relationships between true power (watts), reactive power (VARs), and apparent power (VAs).

FIGURE 27–3 Apparent power, true power, and reactive power.

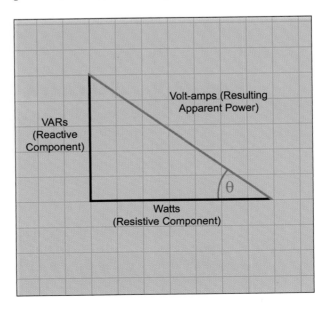

■ APPARENT POWER IN SERIES AND PARALLEL RL CIRCUITS

27.6 Apparent Power Relationships

The apparent power can be calculated by finding the product of the total current flow and the circuit voltage. The relationship for volt-amps, watts, and VARs is the same for a parallel RL circuit as it is for a series RL circuit. Remember that the various types of power add vectorially in any type of circuit. Since true power and reactive power are 90° out of phase with each other, they form a right triangle with apparent power as the hypotenuse (see Figure 27–3).

True power in an RL circuit is consumed by the resistance in the circuit. The reactive power is returned to the source by the collapsing magnetic field around the coil. In both a series and a parallel RL circuit, the power delivered by the source is the apparent power. If the frequency of the parallel RL circuit is increased, the impedance will increase.

27.7 Parallel Circuit Power Factor

Power factor in a parallel RL circuit is the relationship between apparent power and true power. In a parallel circuit, the voltage is the same, but the currents are different. Therefore, the power factor can be calculated by dividing the current flow through the resistive parts by the total circuit current:

$$PF = \frac{I_R}{I_T}$$

Another formula for determining power factor involves resistance and impedance. For example, if you have a circuit impedance of 120 Ω and a circuit resistance of 150 Ω, the power factor can be found with the following formula:

$$PF = \frac{Z}{R}$$

$$PF = \frac{120}{150}$$

$$PF = 0.8, \text{ or } 80\%$$

27.8 Parallel Circuit Angle Theta

When the resistance of a parallel RL circuit is increased, the circuit becomes more inductive because more of the circuit current flows through the inductor. When the inductance of a series RL circuit increases, angle theta (θ) increases. Figure 27–4 shows how the cosine of angle theta is equal to the power factor:

$$\cos \theta = 0.80$$

The angle theta is the angle between the apparent power and true power (see Figure 27–4). In any parallel RL circuit, the voltage across each branch of the parallel circuit is the same.

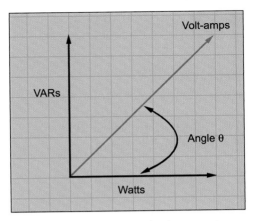

FIGURE 27–4   Angle theta vector diagram.

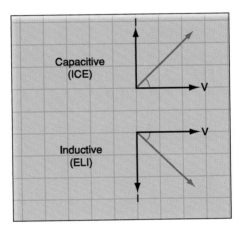

FIGURE 27–5   Voltage and current phase relationships.

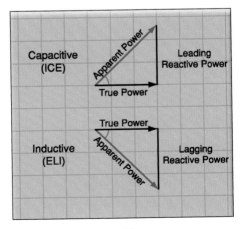

FIGURE 27–6   Phase relationships using power.

27.9 Power Factor: Leading or Lagging

When considering power factor in a circuit, it is necessary to determine whether the power factor is leading or lagging. The concept of leading or lagging must be viewed in terms of the phase angle relationship between the current relative to the voltage.

When the current in a circuit is a result of inductive loads, the voltage leads the current. When the circuit current is a result of capacitive loads, the current leads the voltage (see Figure 27–5).

The next step is to calculate the reactive power for these inductive and capacitive loads. When the angular relationships for capacitive and inductive power are placed on the same diagram (see Figure 27–6), the concept of leading and lagging begins to emerge. Using true power as a reference on the horizontal line and solving for the resultant reactive power found by vectorially adding the power from the inductive and capacitive loads that are 180° out of phase (found on the vertical axis), the location of the resultant reactive power above or below the horizontal reference determines the leading or lagging relationship of the resultant power.

Look at Figure 27–7. If the resultant reactive power is above the horizontal line (positive theta), the power factor is leading. If it is below, the power factor is lagging. Thus, an inductive power factor is referred to as a lagging power factor and a capacitive power factor as a leading power factor. Power factor is never referred to as a negative value but rather as leading or lagging.

■ APPARENT POWER IN SERIES AND PARALLEL RC CIRCUITS

The apparent power of a circuit is calculated by using the total values of voltage and current. For example, in a circuit with a total voltage of 220 V and a total current of 12 A, the following formula can be used:

$$VA = E_T \times T$$
$$VA = 220 \times 12$$
$$VA = 2,640$$

27.10 Power Factor in RC Circuits

Power factor is the ratio between true power and apparent power. It can be calculated by dividing any of the resistive values by its similar total value. For example, given a true power of 1,440 W and the VA calculated previously, we can find the power factor by using the following formula:

$$PF = \frac{P(W)}{VA}$$

$$PF = \frac{1,440}{2,640}$$

$$PF = 0.55 \times 100 = 55\%$$

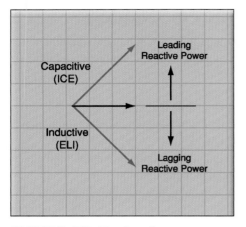

FIGURE 27–7 Leading or lagging power factor.

The power factor of the circuit is the cosine of the phase angle. Since the power factor turned out to be 0.55, angle theta ($\angle \theta$) will be calculated:

$$\cos \theta = PF = 0.55$$
$$\theta = \cos^{-1}(0.55) = 56.6°$$

In this circuit, the current leads the applied voltage by 56.6°. Given the circuit in Figure 27–8, solve for pertinent values.

f_R:

$$f_R = \frac{1}{2\pi\sqrt{LC}}$$

$$f_R = \frac{1}{2 \times 3.14 \times \sqrt{(60 \times 10^{-6}) \times (90 \times 10^{-12})}}$$

$$f_R = \frac{1}{6.28 \times \sqrt{5.4 \times 10^{-15}}}$$

$$f_R = 2,166,922 \text{ Hz}$$

X_L:

$$X_L = 2\pi fL$$
$$X_L = 2 \times 3.14 \times (2,166,922) \times (6 \times 10^{-5})$$
$$X_L = 816.5 \ \Omega$$

X_C:

$$X_C = \frac{1}{2\pi fC}$$

$$X_C = \frac{1}{2 \times 3.14 \times (2,166,922) \times (9 \times 10^{-11})}$$

$$X_C = \frac{1}{2 \times 3.14 \times 1.95 \times 10^{-4}}$$

$$X_C = 816.5 \ \Omega$$

FIGURE 27–8 Series RLC circuit.

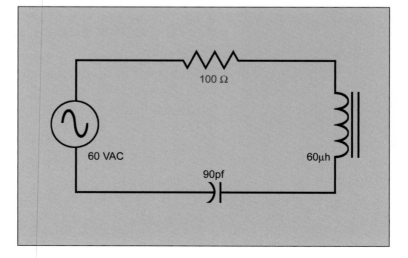

Z:

$$Z = \sqrt{R^2 + (X_L - X_C)^2}$$
$$Z = \sqrt{100^2 + (0)^2}$$
$$Z = \sqrt{10,000}$$
$$Z = 100\ \Omega$$

I_T:

$$I_T = \frac{V}{Z}$$
$$I_T = \frac{60}{100}$$
$$I_T = 0.6\ \text{A}$$

V_R:

$$V_R = I_R \times R$$
$$V_R = .6\ \text{A} \times 100\ \Omega$$
$$V_R = 60\ \text{V}$$

V_L:

$$V_L = I_L \times X_L$$
$$V_L = 0.6 \times 816.5 = 489.9\ \text{V}$$

V_C:

$$V_C = I_C \times X_C$$
$$V_C = 0.6 \times 816.5 = 489.9\ \text{V}$$

Q:

$$Q = \frac{X_L}{R}$$
$$Q = \frac{816.5}{100} = 8.165$$

B: Bandwidth is a frequency measurement. It is the difference between the two frequencies at which the current is at a value of .707 of the maximum current value. It is a term used to describe the rate of increase or decrease in frequency on the way to resonance and is proportional to the Q of the circuit:

$$B = \frac{f_R}{Q}$$
$$B = \frac{2,166,922}{8.165} = 265\ \text{kHz}$$

Theta (θ):

$$\cos\theta = \frac{R}{Z}$$
$$\cos\theta = \frac{100}{100} = 1$$
$$\theta = \cos^{-1}(1) = 0°$$

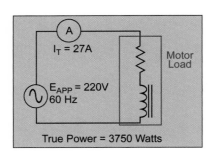

FIGURE 27–9   Motor load.

PA:

$$PA = V_{APP} \times I$$
$$PA = 60 \times .6$$
$$PA = 36 \text{ VA}$$

PT:

$$PT = I^2 \times R$$
$$PT = (.6)^2 \times 100$$
$$PT = 36 \text{ W}$$

PF:

$$PF = \frac{P_T}{P_A}$$
$$PF = \frac{36}{36}$$
$$PF = 100\%$$

■ POWER FACTOR PROBLEM SOLVING

The previous calculations have helped you practice the background calculations necessary to compute the electrical properties associated with power factor. This section applies that knowledge to calculate power factor correction.

Given the following installation, there are a number of questions that can be asked about the circuit relative to power factor. What is the power factor? What is the reactive power? What size of capacitor is necessary to correct the power factor of the circuit? Let's look at the circuit and consider the given information (see Figure 27–9).

Type of load: motors
True power of the motor load: 3,750 watts
Line voltage = 220 VAC
Line current = 27 A
Frequency = 60 Hz

The following procedure allows you to systematically solve for unknown circuit values and determine how to correct the power factor. First, determine the apparent power:

Apparent power = $E_{line} \times I_{tot}$
Apparent power = 220 V × 27 A
Apparent power = 5,940 VA

Now calculate the power factor:

$$PF = \frac{\text{true power}}{\text{apparent power}}$$
$$PF = \frac{3,750 \text{ W}}{5,940 \text{ VA}}$$
$$PF = 0.63$$

FIGURE 27–10   Vector
equivalent of motor.

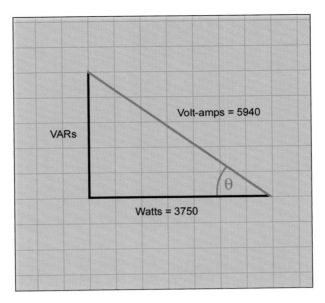

Look at Figure 27–10. What is the phase angle created by this circuit?

$$\theta = \cos^{-1}(0.63) = 50.9°$$

Calculate the reactive power (see Figure 27–10):

$$\mathrm{VARs} = \sqrt{\mathrm{VA}^2 - (\mathrm{W})^2}$$

$$\mathrm{VARs} = \sqrt{(5,940)^2 - (3,750)^2} = 4,607 \ \mathrm{VAR}$$

In looking at this circuit, you can see that power is being wasted by delivering unneeded VAR. To correct this situation, the addition of capacitors in parallel to the motor load will help correct the power factor. Now determine how large of a capacitance must be added to the circuit to correct the power factor.

Since the reactive power is equal to 4,607 VARs, you can calculate the current component of the VARs:

$$\mathrm{VARs} = \mathrm{E}_{\mathrm{app}} \times I_{\mathrm{reactive}}$$

$$I_{\mathrm{reactive}} = \frac{\mathrm{VARs}}{\mathrm{E}_{\mathrm{app}}}$$

$$I_{\mathrm{reactive}} = \frac{4,607}{220} = 20.9 \ \mathrm{A}$$

You can see that 20.9 A of current are being wasted by the circuit because of the inductive reactance of the motor load. To correct this condition, an equal amount of capacitive current needs to be connected to the circuit to cancel out the inductive current. To cancel out the inductive reactive current, a capacitor that provides an equal amount of capacitive current will need to be connected to the circuit, 180° out of phase with the inductive current (see Figure 27–11). These two currents will now cancel each other out. Since the goal is for $I_{\mathrm{L}} = I_{\mathrm{C}}$, $I_{\mathrm{C}} = 20.9 \ \mathrm{A}$.

FIGURE 27–11 Parallel current relationships.

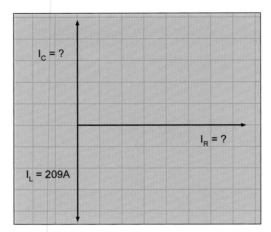

If the capacitor to be connected in parallel will create 20.9 A of capacitive current, calculate the capacitive reactance (X_C) of the capacitor:

$$X_C = \frac{E_{app}}{I_C}$$

$$X_C = \frac{220 \text{ V}}{20.9 \text{ A}}$$

$$X_C = 10.53 \ \Omega$$

This means that 10.52 Ω of capacitive reactance is required to nullify or cancel out the effect of the inductance on the circuit. Given the 10.52 Ω of capacitive reactance, what size of capacitor will be needed to cancel the effect of inductance on the circuit?

$$X_C = \frac{1}{2\pi \, fC} \Rightarrow C = \frac{1}{2\pi \, fX_C}$$

$$C = \frac{1}{6.28 \times 60 \times 10.53}$$

$$C = \frac{1}{3,968} = 252 \ \mu F$$

Based on the information given, by adding a 252.0 μF capacitor in parallel to the load, the total power factor will be corrected to a near-resistive type of circuit. By adding the capacitor, most if not all of the effects of the inductor that resulted in "wasted power" are overcome.

Power factor, although less commonly used, can also be calculated on a series circuit. These calculations, because they are based on the current being the same in a series circuit, will deal primarily with the circuit voltages. Given the circuit shown in Figure 27–12, answer the following questions:

1. What is the power factor?
2. What size of inductor should be placed in series with the capacitor to correct the power factor?

First determine the impedance, Z:

$$Z = \sqrt{R^2 + X_C^2}$$

$$Z = \sqrt{(22)^2 + (10)^2}$$

FIGURE 27–12 Series RC circuit.

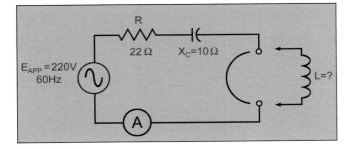

$$Z = \sqrt{484 + 100}$$
$$Z = \sqrt{584}$$
$$Z = 24.17 \ \Omega$$

Now calculate the total current for the circuit:

$$I_T = \frac{E_{APP}}{Z}$$
$$I_T = \frac{220 \text{ V}}{24.17 \ \Omega}$$
$$I_T = 9.1 \text{ A}$$

Now calculate the voltage drop of each of the components in the circuit, recognizing that the current is the same throughout a series circuit:

$$E_R = I \times R$$
$$E_R = 9.1 \text{ A} \times 22 \ \Omega$$
$$E_R = 200.2 \text{ V}$$

Now calculate the voltage drop across the capacitor:

$$E_C = I \times X_C$$
$$E_C = 9.1 \text{ A} \times 10 \ \Omega$$
$$E_C = 91 \text{ V}$$

Now calculate the apparent power and true power:

$$\text{Apparent power} = E_T \times I_{tot}$$
$$= 220 \times 9.1$$
$$= 2{,}002 \text{ VA}$$
$$\text{True power} = E_R \times I_{tot}$$
$$= 200.2 \times 9.1$$
$$= 1{,}821.8 \text{ W}$$

Figure 27–13 shows all the phase relationships that exist in this circuit. The next step is to find the power factor of the circuit:

$$\text{PF} = \frac{\text{W}}{\text{VA}} = \frac{1{,}821.8}{2{,}002} = 0.91$$

Therefore, the power factor expressed as a percentage is 91%. Remember that VAR $= E_C \times I = 91 \times 9.1 = 828.1$.

FIGURE 27-13 Voltage phase relationships.

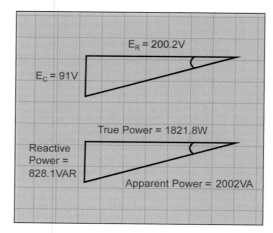

Now that the power factor is known, calculate the inductor that is needed to correct the power factor to 100. What do we know about the reactance of the inductor that will be inserted in the circuit in place of the jumper?

Start by looking at Figure 27–14. To correct the power factor to 100%, whatever size of inductor selected must cancel out the action of the capacitor. In this case, the inductor must cancel out the 91 V dropped across the capacitor; therefore,

$$E_L = E_C$$
$$E_L = 91 \text{ V}$$

Since values for the inductor current and the voltage are known, the reactance of that inductor can be calculated using Ohm's law:

$$X_L = \frac{E_L}{I}$$

$$X_L = \frac{91 \text{ V}}{9.1 \text{ A}} = 10 \text{ }\Omega$$

The required inductive reactance could also have been found by just looking at the reactance of the capacitor. Since a unity power factor will occur when the vector sum of the reactances is equal to zero, the X_L must equal the X_C.

Having the value of the X_L, the value of the inductor in henrys can be calculated as follows:

$$X_L = 2 \times \pi \times f \times L$$

Therefore,

$$L = \frac{X_L}{2 \times \pi \times f}$$

$$L = \frac{10}{6.28 \times 60}$$

$$L = \frac{10}{376.8}$$

$$L = 0.0265 \text{ H, or } 26.5 \text{ mH}$$

FIGURE 27–14   Power factor.

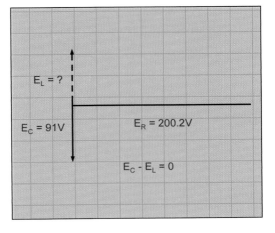

FIGURE 27–15   100% power factor.

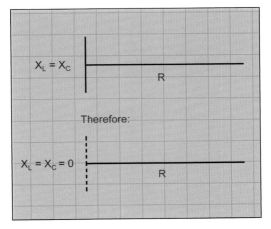

Thus, by adding a 26.5-mH inductor in series with the circuit, the power factor will be corrected to near 100%, as shown in Figure 27–15.

■ SUMMARY

This chapter presents the basic concepts of power factor. Power factor is very important in industrial applications. Electrical power is sold on the basis of power, or watts, consumed. The power company, however, must supply the apparent power. If the plant has a power factor of 60% and is consuming 5 MW of power per hour, the electric company would have to supply 8.33 MVA (5 MW/0.6) per hour. If the power factor were corrected to 95%, the electric company would have to supply only 5.26 MVA per hour to supply the same amount of power to the plant. Many companies are penalized for not correcting the reactive load of their equipment and reducing the apparent power requirement.

Note that the actual generation of the VAR to provide the total apparent power does not cost the power company anything significant. However, transmitting the VAR back and forth from load to generation increases the system losses and causes reductions in system voltage at the load point. Reducing power factor is a cost-saving measure for everyone.

Apparent power is calculated the same way for series or parallel circuits. True power is the watts consumed by the resistive components of the circuit. Volt-ampere-reactive is the reactance components use of power (literally, they store the energy either in magnetic fields or in dielectrics). The combination of these two components (resistance and reactance) gives the apparent power of the circuit measured in volt-amperes.

The power factor of RLC circuits is a ratio of true power to apparent power, written as a percentage. The higher this percentage, the more efficient the circuit in its operation.

■ REVIEW QUESTIONS

1. Define power factor:
   a. As a ratio of power
   b. As a ratio of current (parallel)
   c. As a ratio of voltages (series)
   d. As a ratio of resistance and impedance (series and parallel)
2. If a certain circuit has a 5-Ω inductive reactance, how much capacitive reactance needs to be added to increase the power factor to unity? Why?

3. Most power factor correction is accomplished by placing a capacitor in parallel with the load. Why?
4. Discuss the relationship among VA, VAR, and W.
5. You are working with a customer who is having a low-voltage problem in his power system.

   The system is 220 V single phase and is made up of five 100-W lightbulbs and one 5-kW motor. You measure the total current and find that the load is drawing a total of 40 A. What is wrong, and what could you do to help this customer?

■ PRACTICE PROBLEMS

Practice problems 1 to 7 refer to the following figure:

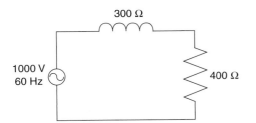

Practice problems 8 to 12 refer to the following figure:

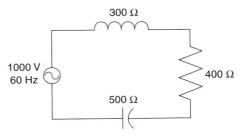

1. What is the current in the circuit?
2. What is the VA?
3. What is the true power?
4. What is the VAR?
5. Do the W plus the VARs equal the VA?
6. Draw a power triangle (vector diagram).
7. What is the power factor and the phase angle?

8. An LCR-A 500-Ω capacitor is added to the previous circuit. Calculate the Z.
9. Find I_T.
10. Find VA, W, and VAR.
11. Draw the power triangle (vector diagram).
12. Find the power factor and angle theta.
13. What size of capacitor should have been used to correct the power factor to 100%?

Index